IN HIS
IMAGE

Also by David M. Rorvik

CHOOSE YOUR BABY'S SEX
(with Landrum B. Shettles, M.D.)

GOOD HOUSEKEEPING
WOMAN'S MEDICAL GUIDE

DECOMPRESSION BABIES
(with O. S. Heyns, M.D.)

THE SEX SURROGATES

AS MAN BECOMES MACHINE: EVOLUTION OF THE CYBORG

BRAVE NEW BABY: PROMISE
AND PERIL OF THE BIOLOGICAL REVOLUTION

YOUR BABY'S SEX: NOW YOU CAN CHOOSE
(with Landrum B. Shettles, M.D.)

IN HIS IMAGE

THE CLONING OF A MAN

David M. Rorvik

J. B. Lippincott Company
Philadelphia and New York

U.S. Library of Congress Cataloging in Publication Data

Rorvik, David M
In his image.

Bibliography: p.
Includes index.
1. Cloning   2. Human reproduction.   I. Title.
QH442.2.R67   612.6   78–5226
ISBN–0–397–01255–1

PUBLISHER'S NOTE

In 1977 David Rorvik visited the J. B. Lippincott New York Trade Division offices and described the extraordinary events recounted in this book. He explained, however, that he had pledged not to reveal to anyone the identities of the other participants, which made it impossible for us to authenticate his story. We deliberated as to whether we should publish it under these circumstances.

Mr. Rorvik was an experienced writer in the field of human biology and a recipient of awards and fellowships for distinction in science writing. The book he proposed to write would inevitably arouse much controversy, but it would explore scientific, social, moral, and religious issues of great import. We believed he would treat these issues in a revealing, responsible manner, and we decided to publish it.

The account that follows is an astonishing one. The author assures us it is true. We do not know. We believe simply that he has written a book which will stimulate interest and debate on issues of the utmost significance for our immediate future.

for my mother
(and other benign predestinators)

CONTENTS

FOREWORD:
A Note on Obfuscation

Owing to the nature and circumstances of the events described herein, I have found it necessary to omit certain details and to alter others in order to protect the identities of those involved. In some instances names, dates, and descriptive details of both person and place are at considerable variance with reality; in other instances these details have been altered little if at all. Throughout I have sought an accommodation of the truth and the protective untruth consistent always with the basic course of events and the essential qualities of character and behavior of the individuals described.

ACKNOWLEDGMENTS

I am indebted to two individuals without whose assistance I might never have found the resolve to thrust myself into this adventure, without whom I might never have found "Darwin," and without whose encouragement I almost certainly would never have completed this book. Regrettably, I cannot identify either of these individuals. I wish also to thank two close personal friends whose counsel and encouragement were of great value: Richard Davis and Andrew R. L. McNaughton.

All is changed, changed utterly,
a terrible beauty is born.
　　　　　—W. B. Yeats
　　　　　　"Easter 1916"

PART I: MORALS

Practiced in private and in secret on individual persons, it will slip imperceptibly into our lives.
—Dr. Robin Hotchkiss
Rockefeller University

If there were any good reason for doing it with a man, it could be done now.
—Dr. Kimball Atwood
University of Illinois

There is nothing to suggest any particular difficulty about accomplishing this in mammals or man, though it will rightly be admired as a technical tour de force when it is first accomplished. It places man on the brink of a major evolutionary perturbation.
—Dr. Joshua Lederberg
Nobel Laureate

CHAPTER 1

It began in September 1973. The phone rang in my cabin on Flathead Lake in the mountains of western Montana. A man who would not give his name said he had obtained my number from a magazine that had published an article I had written. He said that he had read my books and was interested in many of the things I was writing about. He described himself as a fan. I thanked him politely, wondering what he wanted. He said that he was getting up in years, was still single, and lacked an heir.

I was by this time accustomed to receiving calls from individuals who had read things I had written. Most of the callers were women concerned about problems of pregnancy, contraception, infertility, genetics, and other topics I had explored. Usually they wanted more information, occasionally advice. In this instance I guessed my caller was worried that his age might prove an obstacle to conceiving a child. I asked him how old he was. He said he was sixty-seven but still "vigorous." I told him I doubted he had any cause for concern. Many men fathered children at his age and beyond.

Still, he obviously wasn't satisfied. He said he was interested in exploring "all options" and was fascinated by the whole area of genetic engineering. Other than for myself, he said, he had never encountered anyone in his voluminous reading whose interests so closely paralleled his, and he doubted that I would find anyone more receptive to the ideas I had offered in my writing than he was. He suggested that we meet to discuss common interests.

A little annoyed by all this, I said that my time was limited, that I was committed to a number of free-lance assignments. He responded that *he* might have an assignment for me, assuming that by "free-lance" I meant I was available for any bona fide offer. I told him it was unusual for an individual rather than a publication to approach a writer, and then, thinking it might be easier for him to express whatever he had in mind in writing, I suggested he send me a letter.

He told me bluntly that he was a man of wealth and that it would be a simple matter for him to fly to Montana from New York to talk with me in person. As a matter of fact, he could be in Montana on a specific day the following week, as he would then be en route to the West Coast on business. Not the least reserved of individuals, I felt threatened by so much forwardness and blurted out multiple reasons why such a meeting could not take place: I was expecting guests, I might have to leave on assignment, and so forth. I asked again for a letter.

He said he was not ready to commit himself "to print" and asked if he could call me again. I said yes but warned him that we would get no further unless he was willing to be more specific at that time. He closed with something that seemed, on the one hand, like the most embarrassing and fulsome praise and, on the other, almost like a warning or threat. It was difficult to forget.

"Right now," he said, "you may be the most important man in the world. Be careful." I waited for something further, but all he added was "Nice to talk with you," and we hung up.

My reaction to this was mild mystification and impatience. His failure to come to any palpable point left me wondering if he might not be something of a crank. That and his final comment. Still, there was something about the caller's tone, his confidence

—and the fact that he spoke knowledgeably of a number of scientific matters—that was impressive. I was particularly surprised that he seemed to know so much about a series of experiments that had not yet been publicized outside of scientific circles —experiments that would soon launch an international controversy. These were experiments in which the genes of unlike species were spliced together in the laboratory to create new life forms.

I could not, anyway, dismiss the man out of hand. My feeling was that his interest was personal and private, but, given his admittedly vague description of himself as a successful entrepreneur, it was always possible that he envisioned some commercial application of something I had written about. The talk of lack of an heir might simply have been a cover to gain access to other information I might have. But what? On the other hand, if he really was intent upon producing an heir, it seemed that he was not interested in doing so by the standard procedure of sexual conception. What other "options" were there?

One that came to mind almost immediately, since I had written a great deal about it, seemed so improbable at first blush that I thrust it aside while I considered some other possibilities. The man had said not simply that he wanted an heir; he was very specific in saying that he wanted a *male* heir. Perhaps he was interested in some of the things I had written about the sex-selection theories of a doctor at Columbia University. By timing intercourse in relation to the phase of the menstrual cycle of the woman, among other things, the doctor claimed to get good results. But, if the caller had wanted a consultation with the doctor, he could have phoned him directly.

Or perhaps, fearing himself too old or unfit for fatherhood, he hoped to find two eugenically "ideal" specimens as egg and sperm donors and come by an heir either through their actual mating or through the commingling of their germ cells in a test tube. I had written a good deal about such possibilities. My caller had, in fact, mentioned test-tube babies and embryo transplants.

21

Still, my mind kept coming back to the wilder notion.

The following week, on the day the man said he could visit me in Montana, he called again. He hoped I had reconsidered and would agree to meet with him, either in Montana or in San Francisco; it was very important that we get together at once. He said he would pick up my expenses if I would fly to San Francisco.

I asked him to explain what he wanted. And I insisted that he tell me his name. He was evasive. I told him the only idea I had so far was that he wanted an heir, and, if it were a son he wanted, all the information I had on the Columbia doctor's theories were in articles and a book I had written. He nibbled on this a bit—asked me if I really believed the theory. I said it seemed to work and that there was some outside confirmation since I'd first written about it. He said that he *did* want a son, then revised that to, as I recall, "Well, not *exactly* a son."

In my exasperation I think I said I wasn't going to play Twenty Questions and that either he could state clearly what he wanted or forget about it. There was a long silence I remember well, fearing that he might hang up—for my curiosity was now thoroughly aroused—after which he observed that I had written a lot about "cloning," a process by which you could, *without* the union of two sex cells, reproduce a plant, an animal, or theoretically even a human being and do so in such a way that the offspring would be the genetic twin of the organism cloned; that is, its genes and hence its inherited characteristics would be identical to those of its "parent." The clone of a human would, of course, be a baby of the same sex.

Despite the fact that this had been just the "wilder" idea that had occurred to me, I was stunned. In the several seconds in which I could think of nothing to say, my caller said more in one sentence than he had conveyed in many minutes of prior conversation. The terms of his proposal were delivered in one hard-to-digest lump: he would, he said, spend one million dollars and possibly more to attain a clonal reproduction of himself. My job would be to find a doctor or doctors willing to attempt such a task. He felt that, given my interest in this area and given the things I had written, I should be in an excellent position to know who the best candidates would be. He said he had read several of my articles on cloning and genetic engineering in such magazines as *Esquire* and *Look* and had read two books I had written on the subject. We could negotiate a contract, he added, just as soon as I would agree to meet with him in person. If I declined, he would deny having had any conversations with me.

The money, the sheer heft of it, shocked me as much as the proposed goal. In combination, they scared the hell out of me. Cloning was, to put it mildly, a touchy subject in the scientific community—not too far removed from Frankenstein's monster, at least in some people's minds. In my own writing on the subject I had tended to see its potentially positive side. I was intrigued by it—but no more so than had been many reputable scientists, including some Nobel Prize winners.

But for money, I quickly said for the record—in case there was one—that one couldn't buy something like this. It was a naïve statement, I knew, but I felt I had to protect myself at that moment. My caller was obviously prepared for this. He suggested "excitement" as an alternative motivating factor. He asked me not to make up my mind on the spot; could we talk again in a few days? I said yes, provided he would give me his name. I assured him that I would keep his proposal confidential but made it clear that I could not risk any further involvement without some minimal reassurance that the person I was dealing with was all that he said he was. He agreed after some brief hesitation. The name he gave me was oddly anticlimactic. I suppose I had expected the author of so astounding a proposal, particularly one who seemed so protective of his identity, to be, if not a celebrity, at least a person with some sort of recognizable name. I had never heard of the guy.

CHAPTER 2

If I was ruffled by the first call from this man, I was racked by the second. There was an almost frightening discontinuity in the fabric of his conversation; it was as if its form and content were from different dimensions and had never been meant to share the same space. Mad proposals were supposed to be delivered in shrill —even incoherent—cadences. My caller was calm and collected. He spoke in understandable, even commendable English. And that made what he had to say all the more jarring.

For a couple of days I didn't think too clearly about the issues this proposal raised. I alternated between the feeling that the guy was a fraud or somehow out of his head and the notion that he was sane and in earnest. The latter was by far the more troublesome possibility—for then I would have to decide what to do next.

I was five years out of graduate school at Columbia, three years out of *Time* magazine, having opted for the public uncertainties of lone free-lancing over the anonymous securities of group journalism. I'd always cherished my free-lance freedoms, but suddenly I wished I had a science "section" or a senior editor to confer with, somebody or something that could vote or decree or collectively decide on the pros and cons of this matter.

On the other hand, in this period I also found myself aware of the possibility of landing one hell of an assignment, a real change of pace. I could keep my integrity by telling the guy that I doubted whether what he wanted could be achieved but that I'd investigate the situation in depth and present him with the facts. If he was half as successful as he had implied, the fee might be a fat one.

And then I'd drift back to reality. I'd made a successful career for myself in free-lance journalism. Could I afford to jeopardize that by taking off on a wild-goose chase, by becoming involved in what some of my peers might rightly regard as a snow job on an unsuspecting old man? I was annoyed even then that these mysterious calls were making it difficult for me to concentrate on a magazine article that was already overdue.

Finally, in an effort to quiet my mind, I decided to do a little research on my caller. If he was the successful businessman he claimed he was, I should be able to get some information on him.

It seemed to me that a good friend of mine who was an editor on a financial publication in New York should know something about the man. It turned out he did, though not a great deal. Over the phone he confirmed that such a person existed and was, as he put it, "very big" in a particular industry. He checked the reference files and came up with a couple of dated stories on the man. They noted his ability to engineer difficult mergers, his willingness to step on competitive toes when necessary, and his penchant for keeping a low profile.

One unintended effect of my call was to pique the curiosity of my editor friend; he was amazed that so little had been written about a man with so much apparent clout in the business world. He said he might try to do a profile on the man unless I already planned one—he wondered what my interest was. I told him a little about the calls I had received (though not their true intent) and asked him not to talk about them, or to mention me if he did contact the man. I also asked him to call me back with any additional information he might find. He promised he would.

For the purpose of initiating an interior dialogue, I now assumed that my caller was in earnest, though it would be some time before I would, in an operational sense, be completely convinced of that. As the initial shock of his proposal began to wear off, two questions loomed large: *Could* this be done, and, more important, *should* it be done? If the answer to both questions was "Yes," "Probably," or even "Under some circumstances," then I would have to ask myself: Should *I* become involved? If not, I had to wonder if it would be fitting for me to try to prevent others from becoming involved.

I have said that I was, if anything, somewhat favorably dis-
posed toward the prospect of human cloning, provided adequate
safeguards existed and all due precautions against identifiable
areas of abuse were exercised. But all that had been in theory and
on paper. I could not now approach the issue in so academic and
remote a fashion. A man whom I might reasonably assume to be
on the level was actually asking me to act as a go-between, a
courier in a million-dollar plan that could culminate in what
Nobel Prize-winning geneticist Joshua Lederberg had once described as "a major evolutionary perturbation."

Could I afford to be guilty of or party to such a perturbation?
The word struck me as intrinsically disturbing, its four syllables
encompassing the potential for seemingly infinite disorder.

Some writers had envisioned human cloning as the metaphor
of a new age, one in which man literally remakes himself, re-
creates himself this time in his own image; others as the signpost
at the gates of a brave new world in which natural evolution is
no more and the new era of "participatory evolution" reigns
supreme. There were, I knew, in the rapidly unfolding world of
molecular biology, both promises far grander and potential perils
far blacker than those of human cloning; but their meaning was
almost hopelessly disguised to the public by the inaccessible code
of recondite biochemical equations and abstract recombinant possibilities. Cloning, on the other hand, if suddenly thrust upon the
world as a fait accompli, would stand out like a cosmic sore
thumb, it seemed to me, signaling for some "The End" and for
others "The Beginning." Either way it would cause a hell of a
commotion, of that much I was certain.

This might be bad. Mankind, in my view, was already and
increasingly beset by a sense of rootlessness and unreality, proceeding in part from its incremental remove from the natural
air-earth-water bases of life to the synthetic, prepackaged, media-
manipulated illusions of substance that were all part of the now
fading dream of "Progress" and "The Good Life." To some weary
time-travelers, cloning might be a heavy blow, heralding the irre-

26

versible approach—if not the actual realization—of the synthesized, plasticized, carbon-copied Man. To these, the new man, like the new bread—processed, refined, bleached, artificially preserved and fortified, baked to absolute uniformity, and confined in a plastic skin—would be soulless.

And even if I regarded such despair as misplaced, I still had to reckon with its impact, try to gauge how deep it might cut, how the human spirit, already sagging, might decline further under its weight. Even if I regarded as shallow the beliefs of those who would see in clonal humans a threat to God's place in the scheme of things, I could not count myself human, too, and disregard the *feelings* of those thus shaken in their faith.

In addition, I had to confront the possibility that a cloned man might create a biological backlash that would be felt for decades, possibly centuries. Cloning, though in itself novel and fascinating, would have to be accounted trivial in comparison with other developments contemplated by serious and humane scientists in the realm of molecular biology—developments that might make man not only healthier but, ultimately, smarter, better, kinder. There seemed no limit to what might be possible in the wake of a stunning series of breakthroughs that had begun with the decoding of the basic chemical molecules of life. If cloning were to so startle and offend the world, all those glittering hopes of genetic science might be dashed or at least substantially set back, to the detriment of millions.

Last but not least, I had to think about myself, what my involvement in such an undertaking, whether successful or not, would do to my reputation, what it might portend for my own future. I might, it seemed quite likely, be regarded by some as anything but a high-minded idealist who had helped husband a new era. Some might say I had merely pimped and pandered for it, acted as a mercenary go-between, the uncaring croupier in a round of genetic roulette, the dropper, or at least the polisher, of the die in a game of chromosomal craps. It was unpleasant to think about.

Nor was it the only unpleasant possibility for me. Since my caller had hinted that he had no intention of proceeding in the public eye, and since, moreover, I felt that neither would anyone capable of accomplishing this dare brave the recriminations of his or her peers by proceeding aboveboard, I knew that, if my part in this came to light, I would be accused of undermining the traditional scientific ethics of full disclosure I held dear. That would be bad, the more so since I would be guilty. On the other hand, if *I* sought to disclose some of the details but was compromised by the need to protect and conceal my sources—out of respect for journalistic ethics I held equally sacred—I might be doubted, disbelieved, even decried as a fraud. That might be worse.

Still, that "cosmic sore thumb" might, on balance, do more good than harm. If technology could be a demon, it could also be a tonic and a restorative. To many, our first, faltering steps into outer space were proof that our ingenuity and pioneering spirit not only had survived decades of industrial dulling but had prevailed to express themselves in exciting new ways. Some no doubt felt that, in going to the moon and then looking ambitiously beyond that first icy foothold in the once unknowable and forbidding ether, we had overstepped—violated nature and perhaps even challenged God. But more of us, I believed, were lifted by these events—encouraged, if anything, to feel more unified, not only with one another but with the universe itself.

In my view, an even more exciting and meaningful adventure was in store for us on the newly opening frontier of biological inner space. The means were at hand, at last, to launch an exploration into the universe within. The cloning of a man, because its impact would be so immediately dramatic, could make this new adventure accessible to millions who might otherwise understand little of what was transpiring in the molecular world.

It was conceivable, then, that the cloning of a man might not inhibit but actually speed up the important research that was poised, like a rocket on a launching pad, ready to go forward. And if the reaction to what Lederberg had called a "technical tour de force," the cloning of a human being, were to be greeted with mixed reactions, as seemed likely, then one could still argue that this accomplishment had served to focus attention on a field of science which was of the utmost importance to people everywhere but was, until now, largely unknown. Like a red flag, cloning could alert the world to the awesome possibilities that loomed ahead and thus serve as a catalyst for public participation in the life-and-death decisions that might otherwise be left by default to the scientists—men and women who labored for the most part in the interests of humanity but could not be expected to be all-knowing and all-wise.

In that light, my role in all this might be more kindly perceived, it seemed to me. I could argue that through my participation there was hope that these events would be disclosed in some form to the public—that, in short, this exercise would have as its ends something more worthy than the mere secret satisfaction of a single individual. There was an undeniable amount of ego involved in this argument, for I would be saying it was lucky for mankind that I, rather than someone less interested in seeing the public interest served, was involved in this. Still, I found attractive the thought of myself playing as much a watchdog as a promoter or procurer. At least it made it easier for me to contemplate playing any role at all in this project. It was easy to imagine, If I don't do this somebody else will, probably somebody only in it for the money.

The idea of all this was still too tentative, too ethereal for me to weigh other considerations that would later become compelling: What would happen if my caller turned out to be a very clever madman spawning a diabolical power plot? What would happen if something went wrong and this man and his doctors —my God, we—created some sort of monster, misshapen in body, mind, or both? Suppose the product of eventual "success" could not adjust to the uniqueness, the utter strangeness of his being? What if, in my eventual effort to apprise the public of some of what had happened, I should inadvertently reveal the identity of this unique individual, exposing him to what I had no doubt would be the relentless scrutiny of the world?

During those three days I experienced all manner of emotions. It was difficult to keep all this to myself. At times I felt like laughing out loud, telling acquaintances about the crazy fellow who had phoned me. The trouble was, he didn't seem crazy while I was talking to him. And I realized that in any event I was at least partly responsible for his making his proposal. After all, some of my own writings had contributed to his conviction that human cloning was possible and under some circumstances desirable. Was I now going to try to laugh off what I had written, as if the world of books and articles were all make-believe?

CHAPTER 3

When the man called again, I knew at once, even though moments before I had still imagined myself up in the air, that I wanted to explore further the possibility he proposed. He suggested I fly to San Francisco at his expense. I agreed, but emphasized the truth—that he should not construe my willingness to meet with him as any kind of commitment. I said that my mind was far from made up, that I had many doubts about his idea, both ethical and practical, and that, frankly, knowing so little about him, I also had doubts about his suitability for this "distinction." He might, at the end of our meeting, very well find himself out the price of a plane ticket and no closer to getting himself cloned than he had been before. He said he understood.

Before leaving Montana, I took the possibly paranoid precaution of informing a friend of the name of the man I said I was going to "interview." There were moments when my agitated

imagination envisioned all manner of improbable scenarios, all
beginning with my embarking on a journey to meet a complete
stranger and ending with my never being heard from again. The
more practical considerations that I might be wasting my time
and that I should be back in Montana pursuing projects already
in a state of neglect nagged at me all the way to San Francisco,
finally being overtaken by cold, clammy unease as my plane set
down. I was going into this with the same loose-boweled appre-
hension that sometimes preceded other potentially unpleasant or
tense interviews.

As prearranged, I was picked up and chauffeured to the
man's residence in Marin County, a modern stone and glass
structure perched on a hill. It was late afternoon when I was
ushered into a large study and told that my host would join me
in a few moments. The room was lined on one side with books
and, on the other, with guns. He was, when he came in, a much
taller and seemingly younger man than I had expected. I would
have guessed him to be fifty, at the most fifty-five. He was dressed
in a conservative suit and tie. He was wearing wire-rimmed glasses
that seemed tiny on his face. His dark hair was graying and, while
full, was cut short. He appeared, as he had said, vigorous, and
there was an aura of confidence and command about him that I
did not find unattractive.

As he introduced himself he insisted that I call him Max, a
nickname he said he'd had since childhood. I will hereafter so
refer to him, although it was some time before I actually used it
in conversation. He addressed me as David from the outset.

Our initial conversation was small talk, about books, guns,
people we both knew about, and a couple of people we both
admired. It was apparent that he was very well connected and,
from the opulence of his habitat, that he was, as advertised, a man
of means. He mentioned, though not boastfully, other residences
he maintained overseas and in different parts of the country.

Over dinner, served by a very pretty, very silent Oriental
woman, who did not join us, he displayed considerable erudition,

31

with seemingly substantial grasps of politics, religion, economics, philosophy, and literature. His retention of names, dates, places, events, and concepts left me feeling inadequate. I asked him about his education, and he seemed pleased to say that he had completed only six years of elementary school. It appeared that he was very much the self-made man. He seemed intent upon convincing me that he was rational, intelligent, humane; and in that first conversation he made considerable progress on all counts. Still, he seemed reluctant to reveal anything very personal about himself and over dinner deflected one question after another aimed at illuminating his motivation.

After dinner we retired to the study once again. A little sherry was served, and I halfway expected to be offered a cigar until I noticed a No Smoking sign on the front of Max's desk. We had that dislike in common and, it soon developed, were both joggers and near vegetarians.

As I relaxed, I felt emboldened to say that I could not proceed any further unless Max would open up a bit. There were certain things I had to know about his background and his thoughts before I could adequately assess what he was proposing. He would have to share with me some of his innermost feelings or it was senseless to continue—we would get nowhere.

He agreed, but for the first time betrayed some nervousness. All his life, he said, he'd been uncertain about his origins; he was orphaned or abandoned as an infant and passed from one foster home to another. (Later he was to tell me of the fascinating but scanty and not entirely conclusive records about his birth he had dug up.) He said he was sometimes terrified by this void in his life and thought it probably accounted for his exceptional need to exercise as much conscious will over his destiny as he could. He described himself as an overcompensator; he distrusted luck or the leaving of anything to chance.

This trait bore heavily upon the subject of our meeting. Max said that one reason he never had married or had children was that the reproductive process, entailing as it did a chance alignment of hereditary factors in a veritable ocean of genes, went against his grain. He seemed to sense my feeling that this was a bit strange, for he quickly pointed out that he was not alone in believing that reproduction should not be regarded as casually as it had been for centuries. Nobel Prize winners such as Francis Crick believed that having children should be considered as at least as important as driving a car and that prospective parents should have to demonstrate in some way that they were suitable or worthy before being "licensed" to reproduce and assume the great responsibility of parenthood.

He seemed eager if a bit embarrassed to assure me that he not only had nothing against sexual intercourse, but still enjoyed it himself. I thought that he was trying very hard to help me understand him and that this must be difficult for him; he was likely the sort who was used to directing and analyzing others but seldom revealing much of himself. I appreciated his efforts.

Though it would be some time yet before I felt I had a good grasp of the factors that motivated him, his conviction was apparent, even in that first encounter, that the only way he could "die peacefully," a phrase he used more than once, would be to first remake himself, in effect to be born again, and thus give himself the willed and wanted and definite origin that he lacked, or felt he lacked, in his present circumstances.

I said that, although I was fascinated by this dynamic, I was also concerned by it. I thought it would be tragic to try to achieve something that was perceived lacking in oneself through another individual—to try to make that individual, in this instance his clonal offspring, the vehicle by which his own identity, or his perception of it, could be realized. There were serious dangers, it seemed to me, that the offspring might thus be robbed of his own sense of self.

Max said I misunderstood his idea—that rather than do anything to attenuate the identity of "the clone," as we awkwardly referred to this potential person, he would go out of his way to heighten it, for he wished to see how his own life would have been different had he not been hamstrung by his own sense of rootlessness. I countered that many psychologists would argue that his identity problem had in fact made him what he was, the success that he had become, since he was by his own admission the classic overcompensator.

He responded that this might have made him a success in the eyes of the world, but it had never made him happy or secure. And there were practical considerations. He did want an heir, he pointed out, and since he was interested in and believed in the potential goodness of genetic intervention, cloning seemed to him a worthy approach to providing one. I thought I detected a touch of hubris in this, too—the feeling that through the cloning process he might cheat fate, might possibly extend his consciousness beyond the boundaries nature seemed to have dictated. The idea had been put forward that unusual rapport, perhaps even telepathic and prescient, might be experienced by members of a common clone; that is, by the individual and his or her clonal offspring. There had been suggested, as well, the mystical idea that a person's own awareness of the world might somehow survive the death of the body—in the locus of cloned consciousness.

Max wondered what sort of "warp" cloning might put in the flow or "woof" of natural events. He did exhibit a hint of mysticism, it seemed to me, rather than any real sign of megalomania. He said that he never envisioned himself as any sort of cosmic outlaw. The thing that distinguished us from all other species, in his view, was our ability and desire to redefine ourselves constantly, to become new things, and so, to him, cloning was an entirely natural, entirely human possibility.

I played devil's advocate. I said that even if he succeeded he might be badly disappointed. There was no telling how the child might turn out. Max said that he might also be disappointed if he fathered a child by the normal route. This way, at least, the hereditary component would not be in doubt. He had a very strong constitution, had never suffered any major illnesses, and, in terms of his resistance to aging, was regarded as phenomenal by his medical friends. To have a child in the normal way might rob that child of a fundamental physical strength that was almost guaranteed by the cloning option.

We debated for some time the issue of nature versus nurture. Which was the more powerful and determinant, heredity or environment? Max readily conceded the potency of environmental conditioning but insisted that heredity was no niggling force and that its full potential had never been explored.

He showed me several studies and books dealing with identical twins. Obviously he had made a thorough search of the literature on the subject. He spoke of well-documented cases in which identical twins were separated at or near birth and raised in greatly differing environments; yet some of these twins turned out to have gone into precisely the same lines of work, married strikingly similar mates, and in general pursued lives that were remarkably alike. He discussed what some of his authorities called "the simultaneous-death syndrome" which saw twins, separated by vast distances in some cases, dying at the same moment or nearly the same moment, sometimes for no apparent reason.

Although that sort of thing was not the rule, the fact that it happened at all impressed upon Max the potential power of heredity. Members of a common clone, as I had written myself, might prove "even more identical" than identical twins. How they might interrelate no one could be sure. Max asked me if I were not as curious as he was about this. I admitted to being curious—but also frightened.

In further support of the powers of heredity, Max cited the evidence being marshaled by some structural linguists that there is a universal grammar encoded within our genes and that all of us are born with an innate capacity to comprehend any human language. He likened this capacity to a seed waiting to be watered. In line with a growing number of scholars, he was convinced that even morality is genetically encoded—that humans, alone among the animals, have an innate capacity for ethical behavior, though again careful nurturance was required to bring this capacity into full bloom.[1]

Max vaguely implied that *his* nurturance in this respect had

35

been imperfect. He was vowing, I gathered, that no offspring of his would be thus deprived. It was almost as if he were saying he wanted another chance—through his clonal offspring—to be good.

What would the child be told? Max said that he would not try to make the boy believe his mother had died; to do so would defeat his purpose. The child might then, as Max put it, be left with "only half an identity." The idea would be to tell him everything—at the right time, whenever that might be, Max said, indicating he would play it by ear or intuition. This surprised me at first, then seemed like the most reasonable approach. As Max explained it, most children were told stories about their origins that were far more outlandish than cloning. After thinking about it, I had to agree, recalling one inspired modern variant on the old birds-and-bees theme from my own childhood: the parents of a friend of mine told him about Daddy taking his "flashlight" and putting it in Mommy's cave, where Daddy discovered little Jimmy lodged under a rock that only the doctor could remove. At five or six, I think I'd prefer to be told that I had been neatly cloned and cleanly sloughed off someone's body. But then, I said, at twelve or thirteen, with my own flashlight lighting up, I might begin to feel more than a trifle alien. Well, anyone who finds himself in new territory, Max argued, whether it be America in the fifteenth century, or the moon in the twentieth, is bound to feel a bit strange—and probably a bit thrilled.

That, I rejoined, was true, provided they chose to be there. Max answered that none of us, so far as we knew, chose to be born, and yet even among the most miserable there were few who would choose not to have been born.

CHAPTER 4

Max had wanted to talk terms then and there. I refused, saying I needed more time to think; I needed to keep my distance for a day or two before going further. There were serious problems, even provided a doctor who would cooperate could be found. Maybe, for example, human cloning just plain couldn't be done. Max said that wasn't the impression he had got reading what I and several others had written.

He showed me an article in *Atlantic Monthly* by James Watson, who had shared the Nobel Prize for his work in elucidating the structure of DNA. Watson reported that a number of developments were rapidly paving the way for the cloning of man. He said that this event need not be restricted to the superpowers, that lesser countries would soon have the biological technology necessary to achieve it, and that, given the "boring meaninglessness" of so many lives, there would be no lack of volunteers to participate in experiments leading up to it.

Watson feared and opposed human cloning and worried that "a number of biologists and clinicians . . . sensing the potential excitement will move into this area of science."[2]

I agreed that the best minds in science believed that with enough effort cloning could be pulled off, possibly very soon, but there were still other problems. The doctor who would do this, as should be evident from the tone of Watson's piece, would not exactly be an overnight hero. It might not be possible to find anyone willing to risk his or her reputation. Well, Max said, that would be my job—to find out. I said I could just see myself dropping in on serious scientists with the proposal that they toss aside their important work in progress and set themselves immediately to the job of cloning a wealthy gentleman in need of an heir.

Max reminded me that I would not be dropping in with hat in hand—or if it was in hand it would be full of money. With a million dollars my prospects should be able to remain serious and possibly even accomplish a number of nifty things related to their own work on their way to cloning a man. Watson himself had pointed out that vitally needed research breakthroughs would, incidentally, provide the framework for clonal reproduction, and this, Max suggested, could be a major selling point. Only we'd turn things around a bit and argue that arriving at the means for cloning could, fortuitously, provide the framework for breakthroughs in a number of other areas important to the health of man.

But again I resisted any further discussion with Max until I'd reflected in more detail on the issues. The idea of talking dollars and cents at this point was unappealing. I kept thinking, My God, what if someone catches me at this—and with my hand in the cookie jar? I didn't know who that "someone" might be—a paternally scowling Watson, maybe, or one of the sober editors who gave me assignments on sturdy subjects that you could usually talk about in mixed company without any expectation of dirty looks, shock, horror, or amusement. What would some of my sources think—men and women who wouldn't even discuss something like cloning without making a joke or qualifying up to their eyeteeth? I told Max I would stay in the Bay area—at a hotel or with a friend—and get back to him in a day or two.

What I wanted first was a chance to talk with a man I had known while I was a graduate student at Columbia University. He was a particularly thoughtful person with whom I had differed on occasion but whose views I always respected. He had a unique background in medicine, law, and theology and had found the perfect niche for himself as one of the newly emerging "bioethicists," professionals whose specialty was sorting out, in as ordered and humane a way as possible, the ethical ramifications of the new biological research. In a sense, it seemed to me, these bioethicists were the shamans, the priests, of a new age, and perhaps it was only illusion that their version of "right" and "wrong" was any better informed than that espoused by those who still counseled from the cloth. Nonetheless, the only confessional that could possibly suit me just then was one illuminated, at least to a large extent, by the lights of science.

On the phone this man listened patiently while I beat uncomfortably about the bush. Finally, he interrupted with a laugh and a comment to the effect that I sounded like someone who had just killed his mother and was looking for some loophole through which the act might be justified, if not commended. That broke the ice. I asked him to keep my confidence, received his promise, and imparted the whole story of what had transpired so far. I was relieved when he neither laughed nor hung up the phone. He quickly made it clear as we talked, however, that if I expected an easy yes-or-no, right-or-wrong answer I had come to the wrong person. He would guide me to source materials expressing the views of those concerned with this issue, he would enter into dialogue with me, perhaps play devil's advocate—but only I could ultimately decide, he said, whether the proposed project was right or wrong in terms of my own involvement in it.

As my acquaintance led me on an expertly guided tour of my own ethical interior, I was a bit dismayed to find a good deal of this inner space either unoccupied or sparsely furnished. There were philosophical abstractions that were difficult for me to deal with. Concrete realities like abortion proved more useful, and my bioethicist shortly focused on that as a means of getting at some of the issues involved in cloning. I had previously taken a public stand in defense of abortion on demand, but I was no longer very confident of that stand. My more recent reporting on the events of prenatal life had altered many of my previous opinions. I could no longer view with equanimity the casual destruction of a fetus which increasingly—to me, at least—gave every sign of being a living thing with a definite will to survive. I related the horror I had experienced in visiting an abortion clinic where I watched a fetus kick and struggle for life, and my shock and revulsion at news of experiments in which human fetuses whose hearts were beating were decapitated and subjected to other such "experi-

39

mental protocols" in the interest of scientific research, even though some of them, granted, might ultimately result in the alleviation of serious human suffering.

Somehow, I added, groping my way, I was more disheartened by the effect these acts might have on the experimenters—and the rest of humanity—than on these isolated fetuses. Though not a religious person in the conventional sense, I was afraid that these acts were assaults upon all of us, upon whatever it is that is intrinsically good and ordered for us. I was afraid, I said, of the callous disregard for all life that I sensed might cumulatively accrue with the increase of abortion and experimental feticide. The more routine these became, the easier they would be to accept. I was painfully aware that in an overpopulated world there were instances, perhaps millions of them, in which prenatal death might be preferable to lives of endless hunger, disease, and disillusionment, but I wondered if a world which could not provide a better solution to its problems than the mass slaughter of the unborn would be worth living in, in any event.

The bioethicist asked me when I believed life began. I said I honestly hadn't an inkling. He hadn't, either, he said, adding that he wished some of his colleagues who seemed to be quite definite on the subject would share the origins of their seemingly unflappable convictions on this score with the rest of us. Some said the moment of conception, others had no doubt that it was the moment of implantation (when the embryo attaches itself to the lining of the womb and taps into the mother's life-support systems); still others designated the development of the fetal heart and other vital organs as the "moment." Some pointed to the third trimester of pregnancy, and others insisted on the moment of birth. And a few still said it all began with "a twinkle of the eye."

These considerations were important because my worst doubts about cloning concerned the issue of "bench embryos," zygotes (fertilized cells) experimentally conceived in the laboratory and then sacrificed in the process of carrying out some research procedure. In vitro, or test-tube, conceptions were very much in the news, with researchers all over the world attempting to be the first to combine human eggs with sperm in a laboratory container and then implant the resulting embryo into the womb of a woman, perhaps one who could not produce her own eggs, in an effort to create a pregnancy where none could occur before.

A couple of British researchers, heady with early progress toward that goal, spoke confidently of creating many of these "test-tube babies," examining individual cells from them under microscopes in an effort to discover their sex, then implanting those of the sex desired by the parents, while "jettisoning" the others, as one science journal euphemistically put it. In plain English, however, "jettisoning" meant flushing down the drain. The British researchers had already demonstrated that this embryonic sexing technique was feasible.[3] Again, it seemed to me, there was an unconscionable disregard here for what might well be genuine human life—and a disregard exercised in this instance in pursuit of something as relatively trivial as giving prospective parents children of the sex they desired. I wondered if this was substantively different from the "old days" when unwanted girl babies were left on mountaintops to die.

There would no doubt be some bench embryos on the way to cloning the first man, too. The procedure, as it had been worked out in animals, demanded that a body-cell nucleus containing a full set of chromosomes be implanted in an egg cell, the nucleus of which had been destroyed. The resulting embryo would, in turn, be implanted in the uterus of a woman. Many, perhaps dozens, even hundreds of these doctored embryos might have to be created in the laboratory before one could be made to perform in such a way that it would "take" (the egg cell accepting the foreign nucleus), continue to divide, and ultimately produce an identical copy of the cloned individual. In theory, it could work the first time, but it almost certainly would not; almost certainly there would be problems; refinements would be required. Embryos that didn't work out or measure up would be jettisoned.

That was to look at one of the worst faces of it. In another mood, I said, I could easily argue that ends can and often do justify means that sometimes seem unpleasant or even brutal. The bioethicist called me a "situational consequentialist." Was that bad? He said he would call himself the same thing: a person who weighs the pros and cons of each situation and decides whether the good outweighs the bad in each individual case. In this case, for example, an ethical-consequentialist decision would have to take into account the man who wanted to be cloned; his personality, character, and quality would affect not only the resulting clone but ultimately, perhaps, the whole world. It would also involve, he indicated, an assessment of the consequences of this act upon the future of related scientific research.

He had been cautious throughout our conversation but now allowed himself one "opinion," as he carefully labeled it, to the effect that he was not in sympathy with those of his colleagues who maintained that human desire was of no importance or at least insufficient to justify, for example, test-tube fertilization techniques aimed at overcoming otherwise irreversible infertility. Most of those doing the belittling had children of their own, he said, but added that, given their low regard for human desire, he wondered about the psychological sterility of their lives. Desire, he said, making his understatement evident by his tone, was not the least important of all human emotions. Without it we would stagnate, cease to exercise our wills, and thus, in a sense that was very real to him, cease to be human. He wished me good luck, and I knew that he meant it.

I came away from that conversation feeling that, in the view of an individual respected not only by myself but also by many of his peers, an act of human cloning might not necessarily be irresponsible or morally reprehensible. I came away from it, moreover, reflecting on our Promethean universe—a world in which we must exercise the will that distinguishes us as a species, but in order to do so must constantly test the fates—must steal fire from heaven and then hope that molten lava isn't rained down upon us in eternal retaliation.

CHAPTER 5

When Max and I talked again, I warned him that I still hadn't made up my mind—though in truth I realized that in not moving further away from the project I was getting closer to it. Max seemed impatient and insisted that we discuss some of the "logistics" of the matter. I reluctantly agreed, still fearing any talk of money. He said that if I were going to get in any deeper I would have to be convinced that he meant business. He offered to show me financial records—cold cash, if necessary—whatever it would take to prove his willingness to spend a million dollars or more in this pursuit. I told him I didn't need this sort of proof; I was more interested in his background and thought I might do a little further checking on him, if he didn't mind—not that I expected him to emerge as clean as an Eagle Scout. I would be discreet, I said.

He shrugged acquiescence, then asked if I wouldn't accept a fee for my services—which were, specifically, to find the medical-scientific expertise essential to the task. He suggested $10,000 to start and expense money for air travel and the like. He wanted me to be free to travel the world over, if necessary. I said I would consider these offers if I agreed to work with him.

He said he would authorize me to offer a doctor $50,000 just to agree to make the attempt. If the doctor or doctors succeeded, he/they would be given at least $250,000 each. Research expenses could run as high as a million. Also, he could provide medical facilities, lab space, and whatever equipment was needed in a number of countries where he was conducting business at that time. He listed several possibilities, many of them in the Orient.

He then began going into finer details that I felt presumed I was already in the bag, so to speak. I reminded him that I was not, and shifted the conversation to some stipulations of my own.

Provided I did agree to become involved, I said, he would have to comply with several terms, the list of which might grow as I thought about it. He would have to guarantee that I would be allowed to know what was happening throughout the entire effort. I would reserve the right to call a halt at any point if I believed some real abuse of scientific ethics was in process. If the project were thus curtailed, he must make no further effort to perpetuate or duplicate it. If the project succeeded, he must make no effort to repeat it without the unanimous approval of myself and any doctors involved, and he must also agree to let me follow up at regular intervals. By this I meant that he must, in effect, give me visitation rights to see the child at regular intervals for the purpose of reassuring myself, if no one else, that everything had worked out all right.

I explained that, in my view, if this were to be anything other than an exercise in self-gratification, the public must eventually be provided with some knowledge of what had happened. I promised I would never reveal his or the child's identity—stressing that I would take extraordinary precautions on that issue. He acknowledged the "possible merit" of informing the public but said he could not risk exposing the child to the sort of destructive publicity that had accompanied the births and so many growing-up years of so many quintuplets and other multiple-birth children. These were not so much "celebrated," he said, as "made to appear like side-show freaks."

I would never do anything, I said again, that would expose the child, but neither could I proceed with any plan that would leave this effort forever shrouded in complete secrecy. In that case, he said, I would have to be guided by my conscience. Nothing I would write, however, would have his blessing, and he hoped I would reconsider as time went along. Failure to keep identities secure, he said bluntly, would be "unforgivable." I agreed. He said

44

he could accept all of my other terms without difficulty. I told him I needed more time to think and would make a decision as soon as possible.

CHAPTER 6

Maybe it couldn't be done. That, of course, would solve everything. Then I needn't worry any longer over whether it *should* be done. The literature, however, was not very encouraging on this score. For decades, scientific notables had been predicting that clonal or "vegetative" reproduction would one day be practiced on humans. The thing seemed so simple.

Ordinarily, a new human being is created by the union of egg and sperm cells. Each contains a half complement of the necessary number of chromosomes to chart a new life. In combination they provide a full complement, their millions of bits of genetic data intermingling to produce an individual who unites the heredity endowment of his or her two donors. The offspring of this sexual union is thus uniquely his/her own person, and nature, in its tireless quest to keep its options open, gambles on hybrid vigor, on the idea that its seed will, in these new combinations, be enriched and thus eventually go on to bigger and better things. It was a gamble that usually paid off. Or so we all liked to believe —those of us who believed in evolution.

For a long time, however, some scientific seers spoke in their closed circles of another way of making men and women. It was all discussed "in theory," but nobody could think of any reason why it wouldn't work. Except sex cells—the egg in the female and the sperm in the male—nearly every cell in the body comes individually equipped with a full complement of chromosomes, copies of those that combined at the moment sperm and egg—

each with its half complement, as noted above—joined to create that body. In nearly every cell of the body, then, is every bit of information needed to *re-create* that body in its entirety. The main reason these individual cells do not multiply (or try to multiply) into millions of bodies is that almost all of the genetic apparatus of each cell is "switched off" by biochemical suppressors. The only part of the total replicative blueprint that isn't blocked out in a skin cell, for example, is the tiny part that contains instructions on how to make skin. Likewise down the list for bone, hair, heart tissue, nerve tissue, et cetera.

It was hypothesized that if a very dexterous and clever technician, a master microsurgeon, really, could take the nucleus of a body cell—any body cell, practically—and somehow get it undamaged into an egg cell whose nucleus had been removed or inactivated—also without damage—there would be an excellent chance that the total replicative machinery of the body cell would be "switched on" and that it would thus proceed to divide and differentiate, re-creating, in a sense, the individual from whom it had been taken in the first place. If ever a person could be said to be a chip off the old block, it would be this duplicate offspring!

This unique way of making people appealed mightily to those who, however much they might respect evolution, nonetheless thought that nature could use a little help now and then. Such persons spoke optimistically, even wistfully, of cloning great thinkers, scientists, athletes, heroes, entertainers. Later, because many thought that members of a common clone might be able to communicate telepathically (or at least intuitively), owing to their identical neurological schemata, some suggested cloning astronauts, underwater explorers, spies, and others who require or could benefit from unusual rapport in hazardous undertakings. Pessimists, on the other hand, thanked their lucky stars cloning had not been available to Genghis Khan or Hitler and, in their writings and prognostications, they populated the future with ruthless armies of carbon-copied, single-minded doppelgängers under the direction of their despotic donors.

The list of those who had taken cloning seriously was long, and included such distinguished men of science as, for one, J. B. S. Haldane, considered by many one of the most brilliant scientists of the century. It had been his idea to clone individuals with "special effects" such as night vision, lack of pain sense, inability to hear ultrasonic sounds of the sort that might be used in sophisticated weaponry of the near future, dwarfism, which he believed might come in handy in the high gravitational fields of some of the planets we might eventually explore to colonize, and so on. Another was Jean Rostand, the noted French biologist, who thought cloning might be used to convey a sort of serial immortality upon an individual, by replacing worn-out copies of oneself with new ones, indefinitely. More recent arrivals on the list were Nobel Prize winner Dr. Joshua Lederberg, who had suggested that the neurological similarities of body-cell donors and their clonal offspring might serve to bridge the generation gap (donors would be used to teach their clonal young); Dr. Elof Axel Carlson of UCLA, who had proposed cloning the dead "to bring back [historical] individuals of identical genotype" (he believed there might even be enough DNA left "to reconstruct King Tutankhamen from his Egyptian mummy"); Dr. James Danielli, who believed that multiple copies of a single individual placed in different environments would clear up many of the questions related to the nature versus nurture controversy; Dr. Bernard Davis of the Harvard Medical School, who had proposed cloning talented individuals who "might enormously enhance our culture"; and Dr. James Watson, another Nobelist, who saw in human cloning something akin to the collapse of Western civilization but felt nonetheless that it would soon be achieved if we didn't guard against it.

It was all just a lot of blue-sky talk, however, titillating or terrifying, depending upon one's outlook, up until the early 1960s, when Professor F. C. Steward and colleagues at Cornell University took cells from the body of a carrot and placed them in a carefully concocted nutrient bath which contained, among other things, coconut milk. In this solution, something quite remarkable happened. Some of the cells began dividing as if they had been pollinated, though in fact they had not. Some of these switched-on body cells not only divided but finally gave rise to buds and clumps that sprouted roots. In other nutritive mediums they put up green shoots, and, in soil, some of them developed into full-fledged adult carrots, normal in every way.

"We were hardly prepared for the dramatic effects," Dr. Steward said of these history-making experiments. "It was as if the coconut milk medium, thought to be like those in the nutrient fluids of all plant embryos, apparently acted to engage the instructional data in the cell's nucleic acids in such a way that the cell's full potential for protein synthesis and thus for full growth was abruptly unlocked or 'disinhibited.' " So it was, in any event, that the 1902 prophecy of Austrian biologist G. Haberlandt—that clonal reproduction would one day be achieved—was fulfilled at last.

The products of Dr. Steward's ingenuity were called "clones" and the process itself "cloning." The root of this is the Greek word *klon*, which is variously said to mean "twig," "slip," "cutting," or "throng." A clone is now regarded as a cluster of identical cells or organisms—or a single member of such a cluster or throng—all propagated from the same body cell.

It wasn't long until the clones included animals as well as plant members of the living universe. Beginning in the late 1960s and building upon the pioneering work of two American scientists (Dr. R. Briggs and Dr. T. J. King), Dr. J. B. Gurdon of Oxford University repeatedly cloned the African clawed frog. Gurdon first destroyed the nuclei of unfertilized frog eggs with ultraviolet radiation. Then he microsurgically inserted into them the nuclei of frog body cells obtained from such remote places as the lining of the intestines.

Some monitoring mechanism in an egg's cytoplasm (analagous to the white of a chicken's egg) "noticed" that there was a full set of chromosomes in its nucleus. Thereafter this mechanism behaved just as if the egg had come by its full set of chromosomes through fertilization by sperm: it instructed the chromosomes to divide and produce a whole new frog.

In the wake of these startling experiments, noted Cal Tech biologist Dr. Robert L. Sinsheimer predicted in 1968 that it would be possible to clone a human being in ten years. Dr. Lederberg was similarly optimistic—or pessimistic, depending upon how one interpreted his words. Dr. Kimball Atwood, a professor of microbiology at the University of Illinois, said at about the same time that with a "crash program" human cloning could be achieved almost immediately, but that it would in any event likely occur "within a few years."

To many, the accomplishments of Steward, Gurdon, and others showed that the major obstacles to human cloning had already been surmounted. Dr. Willard Gaylin of the Institute of Society, Ethics and the Life Sciences in Hastings, New York, would shortly write in the *New York Times Magazine* that "the Cornell carrot confronts our incredulity. To a scientific mind, the leap from single cell to cloned carrot is greater than the leap from cloned carrot to cloned man."[4]

CHAPTER 7

I spoke with Dr. Kurt Hirschhorn, chief of the division of medical genetics at Mount Sinai School of Medicine in New York. He had served as president of the American Society of Human Geneticists. Did he think the cloning of humans was possible? Not only possible, he said, but probable; in fact, he guessed that it would take place much sooner than people think. Nobody doubted, he said, that human cloning was going to be more difficult than the cloning of carrots and toads, mainly because human eggs must be carried in the womb, not deposited on a rock or dropped into coconut milk. And human eggs are far smaller than frog eggs and more fragile. The biggest challenge for those who would clone a man would be getting the tiny body-cell nucleus into the egg-cell cytoplasm. "But once you get the chromosomes from the body cell in there," Hirschhorn concluded, "there's no reason whatever why the damn thing won't grow just like an ordinary fertilized egg cell."

As for getting the egg cell into the womb so that the clone might be nourished and grow, a technique had been worked out to accomplish this and, in fact, had already been applied to a large number of animals. Reproductive biologists catering to animal breeders, for example, had recently learned how to cause prize cows to "superovulate" by administering various hormones that induced the mass maturation and release of eggs from the ovaries.

The next step was the artificial insemination of these superovulating cows with the sperm of prize bulls, creating in one cow dozens of "superembryos." These were then flushed from the fallopian tubes of the cows and transplanted one by one into the uteri of other, less exalted cows which, as mother surrogates, bore the burdens of pregnancy. In this way a highly valued cow could be made to produce hundreds of calves, instead of the normal seven or eight, in her lifetime—and without ever once having to go through the rigors of pregnancy.[5]

Now, other doctors and researchers, seeking to alleviate the anguish of childless couples, were similarly making advances toward the goal of transplanting human embryos from one woman to another. A dozen or more researchers in several countries were regularly publishing results of their experiments in this effort. It seemed as if a race were on. The first to succeed, it was expected, would reap overnight fame—or notoriety, if something went wrong and one of these test-tube babies emerged mentally or physically defective.

One plan was to remove eggs from the ovaries of otherwise normal women who were infertile only because their fallopian tubes were blocked or had been destroyed or removed. Their eggs would be fertilized in the test tube with the sperm of their husbands and then implanted into their own wombs. Another plan called for taking the egg of one woman, again fertilizing it in a test tube, and then transplanting it into the womb of a second woman who could not produce her own eggs.

Many objected to these ongoing experiments on grounds that, among other things, deformed or retarded babies could result. Others protested that test-tube fertilization was intrinsically immoral, regardless of any possible risks or benefits. Still others feared that such procedures would rapidly lead to a demand by many women for more frivolous and hence more dehumanizing uses of this new technology. Women who simply didn't want to undergo the difficulties of pregnancy, some said, would hire other women to bear their children for them—for a fee. One critic labeled this practice "Wombs for rent." Others said the poor would thus become the literal vessels of the rich. Still, the research continued—and for good and humane reasons, those on the other side insisted. The day was drawing near when it would be possible to successfully implant and even transplant human embryos conceived in glass wombs.

Dr. Watson said that "unexpectedly rapid progress" was being made in the embryo-transfer work and that this progress would inevitably, it seemed, pave the way for human cloning. This goal was made even further accessible by additional breakthroughs—principally those in the area of cell fusion. These made it appear likely, he said, that researchers would soon be able to join body-cell nuclei to egg-cell cytoplasm without even bothering with difficult and risky microsurgery. Chemicals were being found which could induce the two to fuse as if by magic. Viruses were also being isolated that could achieve such fusion.

Indeed, as one of those who opposed cloning, Watson worried that it might be unstoppable, given the fact that the same research advances required to bring it off were also those which would benefit mankind in important and diverse ways. There could be no holding off those advances. Cell-fusion techniques, for example, were promising "one of the best avenues for understanding the genetic basis for cancer." Already malignant cells were being fused with normal cells in an effort to isolate the individual chromosomes that might induce specific forms of cancer. To stop such experiments in order to preclude the possibility of cloning, Watson said, would be regarded as irresponsible by most scientists. Embryo-implantation work that could provide important new insights into contraception and possibly even new solutions to the population problems would similarly be seen as offering benefits that far outweighed the possible risks.

And so I found myself confronted again with the question of *should* rather than *could*. I reread one passage I had written a few years before in which I discussed speculation that, with the advent of artificial wombs—and a couple were then already under development—clonists could forget about mother surrogates and, instead, merely deposit their body-cell nuclei into these mechanical uteri and wait for their copies to emerge nine months later. With the perspective of those few years and the revelations of Watergate washing all around me, I found myself no longer so casual about the possible wonders of ectogenesis—complete test-tube pregnancy and babies "decanted" from bottles—as in Aldous Huxley's possibly prophetic novel *Brave New World:* babies exposed during those all-important, formative nine months to the whims and fancies of contemporary "predestinators" who just might, out there in the real world, have names like Ehrlichman, Haldeman, and Nixon.

Dr. Bentley Glass, who holds the title of Distinguished Professor of Biology at the State University of New York, has stated that ectogenesis will loom as a very real possibility by the end of this century. Mass cloning, Dr. Glass believes, might then be difficult to resist. In an address that was later published, he worried about the temptation to replicate "popular or commanding figures" and envisioned "a concealed clone spread through a population." Popular demand could result in clones of such celebrated individuals as John Kennedy or Dustin Hoffman or Elvis Presley cropping up all over, provided someone had procured some of their body cells—something not terribly difficult to do. A piece of skin could suddenly be worth a fortune on the cloning black market. The mind boggled at the bizarre possibilities. Worse, Glass warned, would be the use of cloning by tyrannical governments. To such regimes, he said, "the production of human clones might seem to have irresistible advantages."[6]

Another noted scientist, the physiologist Lord Rothschild, had recently warned a gathering of scientists that self-centered fanatics in the Hitler mold might be tempted to set up shop privately, making dozens of replicas of themselves on the sly before society could intervene. Dr. Hirschhorn had also warned me about the possibility of do-it-yourself clonists. Lord Rothschild was so concerned on this score that he had proposed the establishment of a Commission for Genetical Control to license clonists —at least implying that in his view cloning might be justified under some circumstances.

Dr. Lederberg, who sometimes seemed to be enamored of the cloning concept, at other times contemptuous of it, and on still other occasions merely bored by it, once expressed the view that where tribalism and racism had prevailed, "clonism" and "clonishness" would assert themselves in the future. The ethnocentrism of clones, embracing perhaps hundreds of thousands of identical individuals, would be formidable, he said, far outstripping mere clannishness or cliquishness.

And each clone, of course, would consist of members of only one sex. They might become so insulated that they would not even mix with the opposite sex, it was suggested. They wouldn't *have* to, at any rate, because they could go on indefinitely reproducing themselves asexually. The female clones, possessed of both wombs and egg cells, would have some advantage at the outset. But the male clones could use slave females or, more likely, artificial wombs. It was not too preposterous, some claimed, to envision a massive "war of the clones" sometime in the future.[7]

53

Though these fears and concerns certainly had to be taken into account, many of them struck me as too fanciful and unfairly pessimistic. I felt, moreover, that some of the critics of cloning were not being entirely forthright in making their arguments. James Watson, for example, took positions and then sometimes seemed to step abruptly aside from them. In his *Atlantic Monthly* piece he contended that the first reaction of most people to the cloning of a human being would be "one of despair." The nature of the child-parent bond, "not to mention everyone's values about the individual's uniqueness, could be changed beyond recognition." To people with "strong religious backgrounds, our most sensible course of action [would be to] de-emphasize [any research that] would circumvent the normal sexual reproductive process." Thus, he added, cell-fusion experiments might no longer be supported by federal funds. "Even more effective would be to take steps to make illegal, or to reaffirm the illegality of any experimental work with human embryos."

This was difficult to deal with because you could not tell whether Watson was *directly* proposing to make illegal or to deny funding to such research efforts. He used as cover, it seemed to me, his statement that this would be the "sensible" thing to do if you were a person of "strong religious background." Was Watson such a person? I too was concerned about the world's reaction to a cloned man, but I was annoyed by what I regarded as Watson's hit-and-run attitude. I wished he had better expressed what *he* thought and, more than that, what he *felt.* I wanted guidance untainted by scientific politics, and I believed that Watson was not being entirely sincere. There was evidence in his piece that he was more interested in taking down a few notches a couple of his admittedly cocky British colleagues active in this field than in formulating any constructive ethical position on the issue.

I thought for a time that my view of Watson's outlook might merely be self-serving. Years later I would find myself both amused and relieved to hear Watson testily describe as "kooks," "shits," and "incompetents"[8] the critics of recombinant DNA experiments, which nearly everyone in the genetic engineering field readily conceded to be potentially far more perilous than embryo transfers or cloning. Watson was deeply involved in the new work but had never had a hand in the embryo or cloning research.

As I reviewed everything I had read or heard on the subject, I saw clearly that nearly everyone who was in a position to know believed that human cloning was a very real possibility and that all that was needed to accomplish it was the encouraging of available talent with money and will. Obviously, someone had to *want* to do this—and probably want it badly—before it could happen. (That someone, I now knew, existed.) But many of those who believed that it could be done also believed strongly that it shouldn't be done—at least not yet.

Still, there were those, and among them were some famous scientists, who had seen fit to point out the positive possibilities of cloning. Lederberg himself had observed that it would make possible almost unrestricted duplication of prize cattle, horses, and other animals important to man. Further, in his words, human cloning would facilitate "the free exchange of organ transplants [between cloned individuals] with no concern for graft rejection"—no concern since the cells of one member of the clone would be identical in every way, and thus entirely histocompatible, with those of any other member. The ability to make valuable duplications of celebrated individuals has already been noted. In addition, some argued, an incalculable amount of basic or pure scientific data would accrue in the process of learning how to clone a man, data related to cell differentiation, cancer, genetics, immunology. Also, as others pointed out, in the wake of successful cloning psychological, neurological, and other consciousness-oriented research might be substantially aided by the availability of multiple identicals.

There were other issues still to be contended with. I had yet to respond in my own mind to the views expressed in the various position papers of those whom my bioethicist friend had referred

55

to as "the heavies," those men of religion and philosophy who were among his colleagues. I was about halfway through my reading in this area when a startling report from New York dramatically catalyzed the making of my decision.

CHAPTER 8

In October, a doctor with whom I had collaborated in the writing of a book and whose work I knew well telephoned me from New York to relate a disastrous sequence of events. This doctor was Landrum B. Shettles. For the past twenty-seven years, he had built a distinguished career for himself as an attending physician, researcher, and professor at Columbia-Presbyterian Medical Center and the Columbia University College of Physicians and Surgeons in New York City. Now, following an explosive incident, he was resigning.

He told me that his superior at Columbia-Presbyterian, Dr. Raymond Vande Wiele, had purposely destroyed a test-tube specimen containing, he said, the sperm and egg of a Florida dentist and his wife, a patient of Dr. William J. Sweeney. Shettles believed that he had achieved a test-tube conception, that the sperm had fertilized the egg, and that the resulting embryo was incubating and could shortly be implanted into the uterus of the wife, whose infertility had been caused by irreparably damaged fallopian tubes.

Shettles was preparing, in cooperation with Dr. Sweeney, a clinical professor of obstetrics and gynecology at Cornell Medical School, to carry out what might have proved a history-making operation when, Shettles averred, he found the specimen had been confiscated, the sterile conditions under which it was being maintained broken, and the entire attempt terminated by Dr.

Vande Wiele. He added that he was called on the carpet and that Vande Wiele "became emotional" and asked him if he were trying to create a "monster." He was then accused by his superior of breaking the law by attempting this procedure and was ordered out of the hospital at once. He noted that the lock on the lab where the eggs were incubated was immediately changed and a guard placed at the door.

In the uproar that ensued, attorneys for both sides finally agreed that Dr. Shettles would resign "voluntarily" and that Dr. Vande Wiele and others at Columbia would refrain from making any further critical remarks about Shettles—that indeed they would refrain from any further comment whatsoever, except to say that Shettles had elected to resign. The Florida couple, Doris Del Zio and her husband, filed a $1.5-million lawsuit against Dr. Vande Wiele in federal court, claiming that termination of the implant procedure without their consent denied them their last opportunity to have a child. (That litigation was, in early 1978, still unresolved.)

I was stunned. For one thing, I was hurt and saddened by what this might do to a man whose work I had written about on several occasions, a man whose humanity and good intentions I could not seriously question. And I could not believe this had happened to a man of Dr. Shettles's stature.

He had received his M.D. and Ph.D. degrees at Johns Hopkins University. His historical microphotographs of the earliest stages of human life, from conception onward, had been included in more than fifty textbooks and displayed in the American Museum of Natural History in New York and the Academy of Sciences in Moscow. Dr. John Rock, one of the principal developers of the birth-control pill, had hailed Shettles's test-tube fertilization of human eggs and his culturing and study of human embryos as "a landmark in our insight into human embryogenesis." Dr. Alan F. Guttmacher, while serving as national president of Planned Parenthood–World Population, said Shettles was endowed with "an ingenious mind and superb technical ability." Dr.

57

Frederick P. Herter, son of the late U.S. Secretary of State and a noted surgeon at Columbia, observed that Shettles's "productivity in the laboratory has been prodigious; he is known the world over for his investigations into the mechanisms of conception and development in early fetal life."

If a giant in the field like Shettles could be shot down for merely making preparations to implant an egg removed from a woman back into that same woman, it was evident that I was flirting with an area of science that was far more controversial than I had ever imagined. Even if many seemingly objective observers felt that a feud involving professional jealousy between Shettles and Vande Wiele had precipitated the event, it was clear that the procedure in question had given Vande Wiele the ammunition he was looking for to "knock off," in the words of one Shettles partisan, "a man whose reputation exceeds his own."

In an interview he granted me in the course of my writing an article that would eventually appear in the *New York Times Magazine*,[9] Vande Wiele claimed that clearance for the procedure had never been obtained by Shettles and that what he did was, in any event, illegal and immoral. For these reasons, he said, he had terminated the procedure. Dr. Shettles, on the other hand, maintained that no clearance was required, since Mrs. Del Zio was Dr. Sweeney's patient and the actual implant operation would be performed at another hospital. Only the in vitro fertilization phase would take place at Columbia.

As for the alleged illegality of the act, Dr. Shettles characterized this as a slander and asserted that no law existed anywhere to prohibit such work. Asked about this, Dr. Vande Wiele pointed to recently published "regulations," as he called them, of the National Institutes of Health.[10] As Shettles had insisted, however, these regulations were still merely suggestions, without the force of law, and at the time they were very tentative suggestions, awaiting the approval or rejection of public review. They had not yet been declared official by NIH.

As I dogged the issue, Dr. Vande Wiele acknowledged that

what Shettles said about the guidelines was true. Still, he would not retract his assertion that the other doctor had acted illegally. Besides, he said, there were still the moral objections, and these were overriding. If there were compelling moral reasons not to engage in embryo transfers and in vitro fertilization, then there would of necessity be equally compelling reasons not to clone. I was still willing to be persuaded.

Dr. Vande Wiele gave me to believe that the bioethicists were pretty well agreed on the immorality of carrying out such procedures until and unless we had more knowledge of the possible consequences. Even then, he said, it might not be considered moral. The crux of the problem, as he saw it, could be summed up in one question: "Who is going to get permission for the fetus?" What he meant was, he said, that if the fetus could not grant permission for its own test-tube conception, then it was immoral to create it—because in the act of creating it there was the possibility of creating something monstrous—retarded or deformed. "This is a very serious consideration," he said.

I agreed that one could not take lightly the possibility of producing damaged babies through laboratory manipulation, that it would indeed be immoral to proceed if the risks seemed substantial. But, I said, Shettles and others believed there was sufficient animal data already attesting to the ability of man to manipulate germinal material and still get healthy offspring. I mentioned experiments in which an entire "herd" of cattle embryos, for example, was transported across the Atlantic incubating in the uterus of a live rabbit (the ideal temporary hostess, it had been discovered) before being extracted and successfully implanted in foster mothers which gave birth to normal calves. Also, mouse embryos had recently been frozen, some at temperatures as low as 452 degrees below zero, thawed, implanted, and brought successfully to term in surrogate mothers. Calves produced in the same manner had exhibited normal growth and sexual viability.[11]

In any event, I added, when fetuses were afflicted with gross abnormalities, nature almost always spontaneously aborted them.

Moreover, some of these abnormalities could now be detected with prenatal diagnostic techniques and instruments. Didn't the risks sometimes justify the ends—alleviating the anguish of those who were childless? Dr. Vande Wiele said the issues were more complex than I might imagine. He suggested that rather than debate the issue further I should read Kass.

CHAPTER 9

Kass, as I recalled from my conversation with the bioethicist I had spoken with, was Dr. Leon Kass, a molecular biologist and bioethicist at St. John's College in Annapolis. It happened that I had already obtained some of his papers, and now I resolved to read them. From what my source had told me, if Dr. Kass was not the principal opponent of embryo transfers and cloning anywhere in the world, he was at least the most vocal. In one of his position papers, Kass described cloning as "the contrived perpetuation into another generation of an already existing genotype." It was his view that the last obstacles to cloning were about to fall by the way. The fusion experiments of Dr. Christopher Graham at Oxford, the in vitro fertilization and embryo-transfer work of others were advancing at such a pace, he said, that "it is reasonable to expect the birth of the first cloned mammal sometime in the next few years."

Again, it *could* be done, but Kass was definite in his view that it shouldn't be done. "Among sensible men," he wrote, "the ability to clone a man would not be sufficient reason for doing so. Indeed, among *sensible* men, there would be no human cloning." But there were apologists and titillators at work, he said, who had constructed a "laundry list" of cloning applications that many might find attractive, if not today, then in the near future:

(1) Replication of individuals of great genius or great beauty to improve the species or to make life more pleasant; (2) replication of the healthy to bypass the risk of genetic disease contained in the lottery of sexual recombination; (3) provision of large sets of genetically identical humans for scientific studies on the relative importance of nature and nurture for various aspects of human performance; (4) provision of a child to an infertile couple; (5) provision of a child with a genotype of one's own choosing—of someone famous, of a departed loved one, of one's own spouse or oneself; (6) control of the sex of future children . . . (7) production of sets of identical persons to perform special occupations in peace and war (not excluding espionage); (8) production of embryonic replicas of each person, to be frozen away until needed as a source of organs for transplant to their genetically identical twin; (9) to beat the Russians and the Chinese—to prevent a "cloning gap."

Kass said that most of the objections he had raised against in vitro fertilization and embryo transplants could be leveled against cloning. "Again, we cannot *ethically* get to know whether or not human cloning is feasible," he stated, "because to attempt cloning would be to place in jeopardy the clonal offspring, the permission of which to do so we cannot obtain." Because it seemed to me that this argument could be applied to reproduction via ordinary sexual intercourse (pregnancy and childbirth are far from risk-free), thus foreclosing, were we to take it seriously, the future of the species, I was relieved to find that Kass had other objections—unique to cloning.

He argued quite validly, I thought, that a cloned individual might suffer a serious identity crisis, finding it difficult to distinguish himself from his donor. He spoke of the "problem of identity faced by identical twins" and suggested that the difficulties encountered by someone who was the "child" or the "father" of his twin might be enormous.

Kass was willing to go one step further. Beyond the possible

61

psychological ill consequences of cloning was another and perhaps more fundamental concern. "Does it make sense," he asked, "to say that each person has a right not to be deliberately denied a unique genotype? Is one inherently injured by having been made the copy of another human being, regardless of which human being?" Kass cited both "modern Western political thought" and the "Judaeo-Christian tradition" as supporters of the idea that each of us has a right to an intrinsic uniqueness in a "special yet equal relationship [with] the Creator."

It seemed to me that even without those traditions, Kass's concern was valid and the question he posed a chilling one. From the beginning I had been uneasy about a clone losing his identity to his donor, even if the donor did not will it. But what Kass was getting at cut even deeper, and I had to admit to myself that the creation of a putative human being devoid of a unique genotype *could* create a "warp," to use Max's word, that would be, in the natural scheme of things, intrinsically malign. What might the untoward consequence of this be? A personality which, even given the best of circumstances, might itself be inherently malign? It was possible.

Kass argued that one's appearance is, "at the very least, symbolic of our individuality." Differences in appearance, he continued, "reinforce (if not make possible) our sense of self, and hence lend support to the feelings of individual worth we seek in ourselves and from others." Membership in a clone with several members he believed would "doubtless threaten one's sense of self; membership in a clone of two might do the same."[12]

Kass had other, in my view less substantive, objections to cloning. He spoke of parental interference in the lives of their clonal offspring. He created as an example a couple who have decided "to clone a Rubinstein." (Famous musicians and mathematicians, he said, were frequently mentioned as prime candidates for cloning.) He envisioned his hypothetical couple forcing their young Arthur to sit down at the piano and play, oblivious to the fact that the environment in which real Rubinstein had emerged could "never be reproduced or even approximated" by them. The child's life would thus be directed along the lines of the parents' desires, all predicated on the accomplishments of the true Rubinstein. Thus any intrinsic potential this child might have had would, Kass argued, be undercut by his elders' unrealistic expectations.

I found some flaws in Kass's reasoning here but tried to keep myself from formulating responses too quickly. I wanted to let the other side speak for itself. Vande Wiele, I felt, had chosen the weakest moral position, the idea that one could not ethically proceed without the fetus's permission, but Kass, from whom this position was apparently derived, had presented stronger arguments, some of which could not be so easily dismissed. I was eager to talk with Shettles, Sweeney, and Mrs. Del Zio herself, if possible. The *Times* assignment on the embryo work—for which I had interviewed Vande Wiele—would afford me the time and opportunity to do so.

CHAPTER 10

Max and I spoke frequently by telephone during this period. He wanted to know whom I was talking to, what I was thinking, when I might give him a decision. He was particularly curious about what was happening at Columbia and what Shettles had in mind to do next. At first, I spoke freely about the Del Zio incident—until a call from my editor friend at the financial publication induced caution. He said he had done some more checking on Max and ticked off a number of things. An orphan, Max never had married. He'd backed political candidates in both parties over the years with no discernible or very interesting pattern, had been involved with some charitable organizations, orphanages, and so on, and had been the target of some litigation ("par for the course"), but had never been convicted of anything. And he had a connection with Columbia University that my friend, knowing about the Shettles affair, thought I might be interested in.

This put my guard up. When Max asked me what Shettles would do next I hesitated, then said he might go to work for a nonprofit institute as director of research. Max thought that perhaps some good would come of his leaving Columbia, that perhaps now he could get on with his research without harassment. He suggested that Shettles would be a natural for the task we were discussing.

In that instant I made a decision I was not to alter. Very simply I lied to Max; I told him I had already talked to Shettles about the matter and that he wanted no part of it, that the idea did not appeal to him at all. Max seemed disappointed. He reminded me of the money and wanted to know if I had made Shettles the offer. Again I lied. I said I had and that he had refused.

Then I said, telling the truth, that Shettles had never exhibited much interest in money, devoting almost all of his life to hospital duties and research rather than to what could certainly have been a very lucrative career in private practice. Max said maybe he would change his mind after the Columbia affair had simmered down. I said he wouldn't. Privately I could only guess that Shettles *might* have been interested in Max's project if his other research could be substantially advanced in the process of carrying out the assigned task, that at any rate he would not have said no on inflexible moral grounds.

But there were two reasons I decided right then never to broach the matter with Shettles. First, it might have seemed as though I were saying, "You have nothing more to lose . . ." and certainly I believed he did. Second, I could not overlook the possibility of a setup. One would have to have been privy to all the details of the Columbia incident, as I had been, to reasonably fear such a thing. With the extreme polarization of views and allegiances that I knew had occurred in the wake of this incident, and even before it to some extent, it was easy for me to imagine someone engaging in an attempt to discredit Shettles decisively. How better to achieve this than by involving him in a scheme to clone a multimillionaire—and then to reveal, moreover, that he was doing it all for money?

The fact that Max was potentially in a position to be chummy with a number of Shettles's enemies inclined me that much more to become cautious, possibly paranoid. I did not believe that a setup was in progress, only that it was possible. I would not take any risk where Shettles's reputation was concerned. After I made it clear that Shettles was not our man, I waited to see if Max wanted out. He didn't, and in subsequent talks, though he continued to be interested in Shettles's work, he never again suggested I approach him. Max's tie with Columbia, I was ultimately satisfied, was purely coincidental and at any rate not a significant one. (Later on I confided these fears to Max and admitted that I had lied about Shettles.)

When I went to talk to Dr. Shettles, it seemed apparent that he had never heard of Max, whose name I mentioned casually in passing. He *had* heard of Kass. His Mississippi origins still evident in his voice, despite decades in the North, Shettles said he guessed Kass was on his tail but that he'd keep on with what he was doing so long as he believed he was doing his patients more good than harm. In the case of Mrs. Del Zio, he said, he was at a loss to understand the offense. The woman produced viable eggs but had no fallopian tubes with which to get them to the womb. She wanted to become pregnant. "If the bridge is out," Shettles asked, "what's wrong with using a helicopter?"

He said that he had read Kass on the issue of embryo work and was at a loss to understand him, too. I had been told by some of Shettles's admirers and detractors alike that, behind the country-boy facade they said he sedulously projected, there lurked one of the cleverest minds in the field of reproductive biology. Shettles was noted for keeping his own counsel. It was impossible to engage him in a discussion on the bioethics of in vitro fertilization, embryo transplants, and cloning. For him, the last word was his patient's. He suggested that if I wanted to talk "philosophy" I get in touch with Dr. Sweeney.

Younger than Dr. Shettles, Dr. Sweeney seemed almost shell-shocked by what had happened. Shettles's fate aside, he had himself been reprimanded by his department head and had found his previously favorable position for advancement jeopardized by his involvement in the Del Zio case. He was concerned, too, about his patient Mrs. Del Zio, whom he reported to be severely depressed by the outcome of the incident. Before that, she had already undergone surgery on several occasions in unsuccessful attempts to overcome her tubal blockage. In one final, desperate effort she had submitted to surgery again, this time to allow for the removal of the eggs that were so shortly to turn up, exposed, on the desk of Raymond Vande Wiele.

Sweeney's views, he told me, were very similar to those espoused by another bioethicist—Dr. Joseph Fletcher—whom I should read as an antidote to Kass, Sweeney suggested. Fletcher's name, like Kass's, had already been placed on my reading list, and Sweeney provided me with additional references for Fletcher. He was at the University of Virginia School of Medicine, heading its department of ethics, and had also served as president of the American Society of Christian Ethics. He and Kass, it seemed, had locked bioethical horns on numerous occasions.

So now I read Fletcher. To the insistence of Kass and theologian Paul Ramsey that only those conceptions that result from heterosexual intercourse are moral, Fletcher argued that laboratory reproduction seemed to him to be, if anything, more "human" because it was more "willed, chosen, purposed and controlled" than the parenthood that was so often the unplanned product of casual intercourse.

It seemed fascinating to me that Dr. Fletcher, who was primarily a man of religious learning, would take a view that one might have expected of a scientist, while Dr. Kass, a biologist, took positions you would have expected of a stern theologian. Kass had argued that it would be immoral to engage in human embryo work or cloning without obtaining a better grasp of the risks involved by doing more animal work; but then he had added that "only humans can provide the test system for fully assessing the risks of using the procedure in humans." With this "Catch-22," Fletcher contended, Kass was taking an antiscience stance, saying that when we don't know everything we are morally obligated to do nothing, consigning us to positions of static passivity.

Fletcher took Kass to task for his "epithets and weighted adjectives" and for reverting to emotional treatment of the issues rather than use analytic reasoning. He chided Kass for characterizing those who would act as mother surrogates in embryo-transfer work as prostitutes. "To many of us," he wrote, "this wish to help others unable to carry a fetus seems generous, even sacrificial."

I found that Fletcher was one of the adherents of the "consequentialist ethic," so recently introduced to me. It was his idea that, in any given situation, a person had a responsibility to reason inductively from the data on the available options to decisions "aimed at maximizing desirable consequences." Those that were desirable were those that many could agree would contribute to human well-being. Thus, Fletcher would not categorically condemn cloning. On the contrary, under some circumstances he could endorse the cloning of humans for any of the nine reasons listed on Kass's laundry list (chapter 9), provided only that "the greatest good for the greatest number were served." For example, he would support the cloning of "top-grade soldiers and scientists [if this were needed] to offset an elitist or tyrannical power plot by other cloners—a truly frightening possibility, but imaginable."

He rejected the idea that cloning and other forms of genetic control would inevitably lead to tyranny. Despots, he said, would no doubt use whatever weapons they had at their command, including cloning and other genetic manipulations if they were available, but, in the absence of these technologies, despots would still find ways of coercing people. "The problem of misuse is political, not biological."

It was Fletcher's view that to accept, as Kass and others

67

seemed to have done, "the invisible hand of blind, natural chance or random nature in genetics" would be a cop-out in much the same way that it would be a cop-out to accept the old theories of feast or famine espoused by nineteenth-century advocates of laissez-faire economics, based on sunspots and the ocean tides. "To be men," Fletcher concluded, "we must be in control. That is the first and last ethical word. For when there is no choice, there is no possibility of ethical action. Whatever we are compelled to do is amoral."[13]

Mrs. Del Zio, who consented to an interview over the phone, clearly agreed with Fletcher. "I can't see why some people believe a baby conceived in this fashion isn't as sacred as a baby conceived in the normal fashion," she said. "There's even more care, more desire, more intent involved here—because so much time, energy, skill, and emotion had to be invested in its conception."

Finally, I came to an essay written by Dr. Daniel Callahan, director of the Institute of Society, Ethics and the Life Sciences in Hastings. He said many of his "intuitions," his "groping thoughts," his "primeval impulses" were with Dr. Kass—he feared that through our genetic gymnastics we might dehumanize ourselves in some fundamental way. But he was disturbed, he said, by Kass's assumption that we already know—even though we seem to be still in the process of self-interpretation—precisely what it is to be human.

Dr. Callahan wished, and wished desperately, it seemed to me, that Kass—or anyone, for that matter—could once and for all prove what the essence of humanity was. Then we would know whether what Callahan called "the sundering of intercourse and procreation" through the likes of cloning was indeed dehumanizing.[14]

I shared that wish. I wondered how long this period of self-interpretation would persist. I wondered if it might never end, wondered if, indeed, the essence of humanity might be *not* to know and thus always feel compelled to test the waters—to steal fire from heaven, as I'd said before, in perpetual hope of discovering what did not exist: the limits of humanness.

CHAPTER 11

There was a danger that I would do mankind harm. I knew that. But this did not convince me that I should not act. Nearly every stride forward had been accompanied by some risk. The risks had to be assessed. Precautions had to be taken. I could insist that a certain amount of animal work be carried out prior to making any attempt to clone a human. I could insist that every prenatal diagnostic tool available be used in an effort to uncover any anomalies. I could insist upon an abortion if it seemed that a "monster"—a baby badly retarded or misshapen—were in the making.

I believed that Fletcher, Callahan, and others had dealt persuasively with many of the objections of Kass, Ramsey, Vande Wiele, and the others on that side. Kass's hypothetical couple with their Arthur Rubinstein clone, for example, seemed to many both to ignore and to assume certain things that could not be ignored or assumed. Unless Kass was positing a totalitarian world in which an elite was free to seize and clone anybody it chose and then farm out the offspring to favored couples, it was difficult for Fletcher and others to conceive of the reality of his scenario. In a world like the one we knew, the Rubinsteins would still be in charge of their own body cells and their clones, if any. Laws that presently protected the rights of parents to their own biological children would surely be extended to clonal offspring as well. If anything, laws would be even more stringent; it was not unlikely that legal and ethical requisites would have to be met by any and all potential clonists.

There was the implication in Kass's writing, as there had

been in Watson's, that once cloning got started every possible misuse of it would ensue. The opposition maintained that, in terms of our actual experience with other breakthroughs, there was no justification for this thinking. Further, as Fletcher had said, evil men would make evil use of essentially neutral technologies, just as good men would make good use of them. And Dr. Lederberg's parallel sentiment was that, if a tyrannical government decided to use cloning and the like to help control a population, "it could not do so without having instituted slavery in the first place." Evil, in short, preceded technology.

Then again, those intent upon cloning a recognized genius such as Einstein would not necessarily force his replica or replicas into the same mold as the model. A more fertile idea would be to provide the opportunity for the new Einstein to choose his own way, making certain only that his environment was an "enriched one." Out of a number of such replicas (and possibly only one) you might reasonably expect a new Einstein to make significant, fresh contributions, if not in physics, then in some other field. It was certainly possible, at any rate, and the benefits accruing from the genius of an Einstein amplified manyfold could not be dismissed lightly in a world beset with ever-increasing complexities and crises.

Arguments related to identity continued to trouble me. But then it could be contended that precisely because, as a member of a clone, you were identical to several others, you might be motivated to achieve superior accomplishments in order to establish individual identity. Or, for that matter, rather than think you had been robbed of uniqueness, you might be more than compensated by an awareness of a stunning "specialness" in a world of largely nonidentical people.

I was still at something of a loss on this issue. I wished that the world's first pair of identical twins had left behind a detailed memoir, although even that could not predict how I would feel toward a clone and its donor. Paul Ramsey had written that he instinctively thought of this relationship as strange or even repellent. I found the potential more fascinating than repellent, but had to wonder whether mere fascination with the question—what will happen, what will this relationship be like?—was ample justification for actually trying to find out. Curiosity, desire, hope. Were those the enemies or merely the adjuncts of reason? Whatever the case, they were all bound to play into my decision.

On the question of what a cloned human would do to society and science—and to my reputation, puny by comparison to those other concerns but still of great importance to me—I could only hope, and grasp at what straws were within reach. A quote from philosopher Alfred North Whitehead, for example, was vaguely comforting. "It is the first step in wisdom," he once said, "to recognize that the major advances in civilization are processes which all but wreck the society in which they occur."

Also comforting was something Willard Gaylin had written (in his *Times* piece mentioned earlier). Though it had discouragingly been entitled "Frankenstein Myth Becomes a Reality," it contained this hopeful passage: "Cloning commands our attention more because it dramatizes the developing issues in bioethics than because of its potential threat to our way of life. Many biologists, ethicists and social scientists see it not as a pressing problem but a metaphoric device serving to focus attention on identical problems that arise from less dramatic [but, he implied, potentially more dangerous] forms of genetic engineering and that might slip into public use, protected from public debate by the incremental nature of the changes they impose." It was nicer to think, at any rate, that I might be partially responsible for a potentially useful metaphor than for Lederberg's possibly awful "perturbation."

Granting the sincerity and wisdom of those with often radically disparate views on these matters, I could not delude myself into thinking that a decision based on clear-cut principles of reason and morality was possible. Nor could there be any tallying up of debate points, followed by the designation of a winner. There was no winner, as far as I could see. Finally, in the absence of any legible signpost indicating authoritatively which way I should go, I decided, with doubts and hopes intact, to proceed in the direction in which, no doubt, my own heredity and environment had long inclined me. I could always, I said to myself, turn tail and run like hell if I had to.

CHAPTER 12

Max was delighted that I had decided to "throw in with the devil," as he none too charmingly put it. He preferred a verbal agreement to any sort of written compact, and that suited me fine, too. My job was to find the medical/research talent necessary to undertake and, we hoped, accomplish the cloning task. In return, Max promised all I had asked for earlier—the right to call a halt if I felt it necessary, the right to be apprised of all that was happening, visiting privileges if we succeeded, and so on.

I agreed to accept expenses but decided against taking any fee. I was still paranoid over the Shettles incident and worried in any event that my objectivity might be badly compromised if I were to accept money for my part in this. I wanted to retain my independence and remain a reporter, not become an employee. I felt that in the long run I would be amply compensated by this opportunity to witness firsthand events that might prove historic.

There were other problems related to my accepting money from Max, if I were going to write about this project. Because he couldn't be linked to me by so much as a single check from either him or one of his business interests, payments would have to be laundered or perhaps made in cash. The legality of any arrangement would certainly be in doubt if I failed to report these payments as income. If the payments were substantial, I would have to identify their source. To conceal large sums of money would put me outside the law, and since I had been doing considerably better than merely keeping body and soul together with my free-lancing and book assignments, I could well do without such risks.

Max would clearly have preferred to have me in his pocket; I never doubted that for a moment. In fact, he openly expressed his fear that as an independent agent I would be tempted to "sensationalize" the story I might ultimately write. This was a specious argument, as I pointed out to him; if I were that unscrupulous I would simply ask him for a hundred grand and promise to keep my mouth shut. Not one to give up easily, he said he would pay that and more if I succeeded in finding someone who would actually accomplish what he wanted done—provided I *would* agree to keep my mouth shut. I declined.

Even if the project should fail, I reasoned, it would nonetheless provide my expenses and further open up for me an area of research about which I was often writing anyway. I had no intention of suspending my free-lance career during the time I would be working on Max's project; I was not about to put all my eggs in one basket, especially one as controversial as this.

The way things stood, neither Max nor I had any legal hold on the other. Still, I felt that my position was strong. With his lifelong penchant for conducting his affairs in private, he would fear public exposure, particularly in so personal a matter as this, far more than he would worry about any legal action per se I might bring against him if there were to be some breach of agreement. For his part, Max had only my word that I would not reveal his identity, but he had made his considerable power evident to me in a number of ways. Were I to expose him, he could no doubt take steps to make my part in this project appear less than high-minded. And though he never threatened me—except to say on that one occasion that my revealing his identity would be "unforgivable"—I knew enough about him at this point to appreciate fully that he was not a man one might cross uneventfully.

73

And now for my job—the talent hunt. It seemed a daunting prospect at the outset—so much so that I had to wonder if all my anguish hadn't been for nothing. I pored over *Biological Abstracts* and *Index Medicus* for days, then for even more days looked up and read the articles that might lead me to one or more of those researchers not only potentially capable of what was wanted but willing to get involved. The "willing" part was the real obstacle. It is very difficult to find clues to being willing couched in the scientific jargon of cell biologists and experimental embryologists.

There were other difficulties. It was hard for me to conceive of myself in this new role. The nature of journalism had changed remarkably over the years. The old standards of objectivity, in which the reporter dutifully listed the who, what, why, where, and when of an event and kept his or her impressions and feelings entirely out of the prose, had been found wanting. "Interpretive" journalism, it was believed by many—and their arguments were often persuasive—was needed to give the "truth" three dimensions. But I was troubled, because at the leading edge of this new perspective were those who often seemed to make rather than report the news.

I found myself uneasy at the prospect of falling into that category. Rather than merely react to a news event, it seemed now that I would help create one, *then* report it. This had disturbing ramifications. Getting others involved and then stepping back under the umbrella of a reporter's immunity would, I thought, make me feel a bit like an agent provocateur. I kept remembering an unpleasant incident from my undergraduate days.

I was editor of a student daily newspaper when an earnest young man who was a member of the Young Socialists or some similar left-leaning organization approached me to say that he was thinking of burning his draft card in public during the scheduled address of a well-known right-wing political figure. Draft-card burnings were, in those early days of the Vietnam war, still explosive news. I saw a hot story and, without ever saying, "Yes, do it," encouraged the fellow in subtle ways so that he gathered enough resolve to go ahead.

I took care to make sure I was sitting right behind him at the address—thereby not only reminding him of his proposed action, which took place, but also thereby being in an excellent position to photograph and report the incident in end-of-the-world type in the following day's edition of the paper. I played the story so big that the regular news media couldn't ignore it— and neither could the FBI.

The student shortly came to me in tears; his life was in a shambles, and even his family had rejected him as a traitor. He was leaving his job, his wife, and his college career and striking out for a new start in Canada. What shocked me most was that he had come to me with all of this, as if I were the last and only friend he had. But if the impact of my role in this "news" event was lost on him, it fell very heavily, at that moment, upon me.

Still, on those days when the excitement of Max's project prevailed, I reasoned that in the adult world anyone smart enough to bring something like this off and willing to take perhaps a million dollars for it could look out for himself. Or herself. There were, I soon discovered in the cross-indexed world of those engaged in "Reproduction, Asexual," "Genetics, Human," "Clone Cells," "Micromanipulation," "Genetic Intervention," and the like a good many enterprising women. They seemed every bit as much at home as the men in the pages of such journals as *Methods in Cell Biology, Journal of Embryology and Experimental Morphology,* and dozens of others.

I constructed from my reading and my conversations with various sources, none of whom was apprised of my real interest, a list with fourteen names on it. Eleven men and three women —many from outside the United States. Most of the names, in fact, were British. This came as no surprise, since I had known for a long time that the British were preeminent in the realm of cell biology and experimental embryology.

The list was not the product of the scientific method, to understate considerably. Italians, for example, and there were some promising ones, were eliminated from it or accorded low priority on the assumption that their probable Catholicism would

75

predispose them unfavorably toward the project. I recalled the uproar in the early 1960s that had greeted Italian scientist Daniele Petrucci's announcement in Bologna that he had produced a human test-tube embryo and grown it, he said, for nearly two months in a laboratory womb. Some of Petrucci's countrymen demanded that he be tried for "murder," since he eventually terminated his experiments, i.e., his embryos. He had been intent, he said, only upon obtaining embryonic tissue for use in important transplantation studies. Petrucci, himself Catholic, had no doubt whom the Pope had in mind when he soon condemned those scientists who "took the Lord's work in their own hands." Petrucci avowed in the wake of this storm that his hands would henceforth be clean, that he would never again conceive or grow embryos in the lab, though, according to some reports, he subsequently helped the Soviets do so.[15]

There were others I eliminated because they were on record as opposing cloning or related research. Some others were ruled out as prospects despite their excellent credentials for the job on the basis of positions or personalities that seemed too forbidding. A man or woman very much in the public eye could scarcely indulge in intensive and, in all likelihood, entirely secret research.

If I was biased against Italians, so was I at least somewhat disinclined against Americans. The idea of finding someone removed from at least the geographical dimension of my present reality seemed somehow reassuring—safer, perhaps, with less opportunity for exposure by the vigilant media here. The ideal candidate would be single and without strong family ties, perhaps recently retired or in private research, someone able to take an extended leave or sabbatical.

The problem here was that, while it was relatively easy for a reporter to compile a fairly accurate dossier on a United States citizen, it was not so easy to check out individuals dispersed halfway around the world. Acquaintances who had helped me on other stories overseas proved useful in some instances. In a few others I simply took to the phone and called my prospects, shamelessly posing under an assumed name as one interested in writing articles about them and their work. Those who didn't speak English—or spoke it poorly—either got moved off or to the bottom of my list, which was under constant revision for some time.

Max offered to provide me with interpreters, but I declined. For one thing, with an interpreter filtering everything I said, I would have to be even more circumspect. For another it seemed to me that I had the obligation not only to protect Max's identity from the prospects but that I must also protect the prospects from him. Partly, I feared that if Max had access to my list and I decided, for one reason or another, to withdraw at some point, he would have less trouble going on without me.

By early 1974, only a couple of months into my search, I thought I had found my man. He was American and had excellent credentials and, by all accounts, considerable ingenuity. And he was, at the moment, between jobs and eager to expand his research. I had interviewed him on a previous occasion and believed from various things he had said and written that he might be receptive. It would be easy for me to talk to him and, I would hope, not tip my hand unless he seemed genuinely interested.

We talked in person in his office. We discussed the embryo work for several minutes, and then I asked him if his views on that topic, which were generally favorable, extended to such things as cloning. Did he regard cloning as immoral in any way? He didn't, he said, adding that even if he were a religious person he doubted he would find cause to oppose it morally. It seemed to him that nothing new was being created. It was "God's DNA, any way you slice it."

Could it be done? Yes, he thought so; the egg could undergo remarkable manipulation and remain viable. "The one thing Mother Nature wants more than anything else," he said, "is a baby." It seemed she didn't care how she came by it—she was willing to make all sorts of exceptions to standard practice on the way to attaining the desired goal.

I asked him how soon he thought it could be done—if

someone really set his mind to it. He said it would depend upon many things—resources, in particular. I felt emboldened to ask how long if someone were willing to spend a million dollars or more. The man laughed and said that in that case he'd have it done in nine months.

When I pressed the issue, he said he couldn't guess, that nobody had ever tried human nuclear transfers, as far as he knew. It might prove very difficult, but even then somebody might luck out early in the effort. Or it might prove astonishingly simple.

I asked him what *he* would do if someone offered to put a million dollars at his disposal if he would undertake the project. He joked that he'd guess Christmas had come early. I eased in a little closer until he asked me if I were serious. I decided to take the plunge: I told him I knew it sounded crazy but that a man whose sanity I could vouch for was willing to spend that much. I said I hadn't made up my mind about the wisdom of it all as yet; I was merely floating a trial balloon.

He wanted to know who the man was. I said I couldn't disclose that until he had expressed a strong interest. He said he would think about it. In the meantime we both agreed to keep the matter to ourselves. I left the meeting elated.

Encouraging developments in what one source had called "the embryo-transplant sweepstakes" added to my feeling at this time that our project might succeed. Dr. Patrick Steptoe of Oldham General Hospital in Lancashire, England, and Dr. R. G. Edwards, a physiologist at Cambridge University, widely regarded as the front runners in the race, had published a number of reports claiming success with in vitro fertilization and human-embryo culturing. A team of Australians claimed already to have achieved a pregnancy following embryo transplantation, but a pregnancy soon aborted by an accident. Shettles declared that he, too, had achieved a pregnancy with one of his transplanted embryos—but in this instance the embryo had been transplanted to a woman scheduled for a hysterectomy.

Shettles and a collaborator, Dr. Wayne Decker, were drawing up a "transplant protocol," to be submitted to a major New York hospital, the first step in what they hoped would eventually make the operations routine.

I was encouraged, too, that more researchers were going on the record at this time with their embryo work. The Shettles incident at Columbia had not yet received any major publicity, and this, I felt, abetted me. (A news blackout of the incident had been part of the agreement under which he resigned.) I feared that news of Shettles's resignation and the circumstances under which it occurred might have a chilling effect on my prospects, and in my conversation with the doctor who had promised to think about my proposal I had, somewhat guiltily, sidestepped any reference to the incident.

Word of mouth had spread the news, however, because when we talked again my doctor prospect asked me if I had heard what had happened at Columbia. I said that I'd heard a few details. He commiserated on Shettles's behalf but said that he didn't believe he was cut out for this sort of thing. He'd be damned if he did everything on the record and aboveboard and damned if he didn't—and got caught. He said he was astonished to discover that this was a risk he wasn't willing to take, "even for a possible million dollars."

It seemed gauche, but I asked him if he could recommend anyone else. He hesitated for what seemed a very long time and then, on condition that I would not say he'd sent me, he gave me a name I had never heard before. He filled me in briefly; then we both agreed to forget we'd ever discussed the matter. A man of integrity, he would keep his word, I was certain.

CHAPTER 13

Even before we'd found him, Max called him "Darwin." He was the man we both hoped would, with more than a million dollars at his disposal, clone a human being. Because almost any specific about Darwin could contribute to his identification, I have decided that I must say almost nothing about the details of his life, his position, his background, or his nationality. I will say that he does speak English, that he is middle-aged, tends slightly to overweight, and is a bit flamboyant in his apparel. And while I had overlooked him in my initial canvass of the possibilities, I subsequently found that he had published several scientific papers (which, however, had I spotted them earlier, would probably have seemed to me to be of only marginal interest) and was well regarded by those few who knew his work.

The worst that I could say of him was that he was indecisive. This reflected a healthy, perhaps too healthy, regard for his own hide. I guessed that he was interested in the money and the opportunity to further his own research that the money would afford, though I purposely avoided knowing the details of his financial accord with Max. Darwin seemed torn throughout this effort between the attainment of whatever reward in all this might be his and the ever-present, although variably pronounced, conviction that he was about to be, or already had been, shorn of his scientific virginity. Worse, it often seemed that he feared people could tell just by looking at him. In between these episodes of insecurity he further unsettled me, at least, by exuding almost boastful confidence.

Darwin and Max met privately; then all three of us met together. This time the place was Max's office in Manhattan, a large room notable mainly for its emptiness. Though the glass and steel obelisk of which it was a part looked impressive, Max referred to it as a "dump." The building, like so many around it, he complained, was a cut-rate skyscraper that would fall over in the first big wind. He maintained only a token presence in the building, though the offices below were occupied mainly by people who worked for him. Most of them, he boasted, knew him neither by name nor reputation.

Darwin, who had evinced some doubts about Max when I had first mentioned him, now seemed clearly impressed. The two men were quite a contrast. Max was tall, cool, and collected. Darwin was short, a bit dumpy despite his expensive clothes, and prone to fidget. Even here there was a No Smoking sign, but Max announced it suspended in Darwin's honor. Still, the doctor merely kept passing his cigarettes from one hand to another.

Darwin was not so nervous, however, that he feared speaking his mind. He was particularly displeased with the possibility that I might someday write about this adventure. When I reiterated that this was indeed my intention, he became almost testy. One word led to another, and soon Darwin was flatly refusing to proceed under what he termed the threat of exposure. I assured him again that I would protect his identity. He insisted that I might slip and announced that he would pull out if I didn't promise firmly never to write a word on the subject. I refused. He would not, then, he said, discuss the matter any further.

Both Max and I were flabbergasted when Darwin stood up, took his coat, and said he was returning to his hotel room. I started to say something, but Max signaled me to be quiet. After Darwin had left, he said it was best to do nothing for a few hours. He felt Darwin would cool off and reconsider. And so I returned to *my* hotel room. Then for two days the three of us talked to each other—on the phone. Max would phone Darwin, then phone me and tell me what was said. Darwin finally understood that Max's commitment to me was ironclad—and that was that. Darwin didn't like it, but he obviously liked the terms Max was offering him well enough to live with the threat of a journalist hanging around, after all.

Having introduced Darwin to Max, I think Darwin felt that

I had expended my usefulness and should have no further say in things. I could understand his feelings but was naturally disinclined to share them. On the other hand, there were times when he seemed to want to be very nice to me; I usually thought, perhaps unfairly, that those were the times when he realized it might be better to have me on the team than off and "blabbing," —a memorable gerund he had used to express his fear of me in that first discussion.

When we met in the same office again, there were a number of problems to be solved, more of Max's logistics. Where, first of all, would the research be conducted? Darwin and Max agreed that if it could be done, or most of it done, outside the United States, this would simplify things. I agreed, too, but being determined to play the devil's advocate from the start in the hope of fulfilling some useful function other than mere spectator at a covert scientific spectacle, I added that working outside the States would certainly make it easier to proceed "Russianlike." This was a term one of my sources for my *Times* piece on the embryo work had used to describe the undercover tactics of some of his competitors in the embryo-transplant race.

Darwin only nodded, then pointed out that the rules of informed consent (making sure experimental subjects understand what is being done to them and give their permission) were much more relaxed in many other countries. I winced a bit at that, and Darwin got my point, but argued that the risks were minimal and that eggs would be obtained only from women already scheduled for laparotomy (abdominal surgery)—or via minor surgical manipulations.

Max proposed a half-dozen places where he thought the research could be conducted. These ranged from South America to the Orient. The possibilities were finally narrowed to two sites, both in places where Max already had considerable interests. Both had medical facilities Max had helped fund. Both were out of the way, yet were served by air and were readily accessible from several major cities. Equipment, Max said, could be easily and unobtrusively shipped in as needed with the large and regular airfreight consignments of one of his business concerns. Max would personally show Darwin both facilities, a list of needed materials would be compiled, and research would, he hoped, begin within a month—in early spring.

In late April 1974, I received word that a site had been selected and that Darwin's lab was already being equipped. Some research assistants had been obtained, though they were unaware of the true nature of the project. Darwin had found a way of delegating some of the lab work without identifying his goal.

In July all work was abruptly suspended. Darwin was quitting, he said—"amazed," to use his word, that he had ever become involved in the first place. Once again, Max and I were flabbergasted.

The trouble had begun just as abruptly. Word spread quickly around the world; it was on all three television news networks. What had happened was this: Dr. Douglas Bevis of Leeds University in England, scheduled to address a meeting of the British Medical Association, had first passed out to the assembled researchers and reporters a press release which stated unequivocally that three human embryos conceived in test tubes had been successfully implanted in the wombs of three women. All three had given birth to healthy babies. The feat that Shettles had so nearly brought off before he was shot down had now, it was claimed, been accomplished successfully for the first time in history.

All three women, the release stated, were regarded as hopelessly infertile due to damaged or absent fallopian tubes. Researchers had surgically removed eggs from these women, fertilized them in the test tube with the sperm of their husbands, and then reimplanted them into the women. Out of more than thirty such attempts, Dr. Bevis said, three had been crowned with success, and the three babies now ranged in age from twelve to eighteen months. All were said to be developing normally.

A considerable controversy—and one that literally panicked

83

Darwin—arose in the wake of this announcement. Because Bevis had worked in secret and was making his claims without scientific documentation, his scientific peers and especially his competitors were bitterly attacking him. Darwin apparently saw in all this a pale preview of what might lie in store for himself, should his part in the cloning work ever be revealed.

CHAPTER 14

I knew Bevis's work. I also knew that he was the sort of man who is usually described as "noted" or "distinguished." He was known in particular for his pioneering contributions to the solution of the serious Rh disease in which the mother's antibodies destroy the fetus's red blood cells, killing the baby or leaving it so weak with anemia that it would die soon after birth. His work had resulted in the death rate from Rh incompatibility being cut in many hospitals by more than 50 percent.

But eminent as his career had been, Dr. Bevis was now, with his announcement of successful human-embryo implants, damned and denigrated by a sizable segment of his peers, not to mention those who regarded all such experimentation as outside the proper province of mankind.

It was true, as many said, that in making his announcement Dr. Bevis behaved in a most unorthodox fashion. First, his press release about the test-tube babies was not, as one might have expected, enlarged upon in the scientific address he thereafter delivered. He made not a single reference to the matters covered in that press release, an omission that left his colleagues—and the press—breathless with wonder and frustration.

Naturally, as soon as he concluded his talk he was beset by a barrage of questions. Who had done this? Where had it taken place? Who were the women? Where were the babies now? Why hadn't those involved published their results? What precautions had been taken to guard against deformities?

Dr. Bevis claimed at first that he was not a member of the team that had achieved these firsts. Later he admitted that he was, and under more intensive questioning yet, he shortly said that there was no research team and that he alone had carried out these procedures.

The confusion of all that aside, everyone was by this time condemning him for not presenting his data in an orderly fashion, with all the details spelled out, preferably in an established medical journal. Still, most of Bevis's competitors, when I subsequently interviewed them, privately acknowledged that the Leeds researcher is a man equally of integrity and ingenuity and that, if he said he had accomplished these things, then no doubt he had. It was the secrecy they couldn't tolerate.

Bevis continued to insist that he could not reveal the identities of the women involved. To do so, he said, would be a gross invasion of their privacy. The parents did not want their offspring turned into overnight celebrities, to be hounded by the press or harassed by those who regarded such doings as immoral. Thus, he said, he would not, indeed, could not ethically document this work in the way that he had always so scrupulously and carefully documented his research in the past. Obviously, his unusual and incomplete announcement at the meeting of the British Medical Association had been a trial balloon by which he hoped to gauge public reaction to the work.

That reaction soon landed him in the hospital—sick with disillusionment, associates said, as well as exhaustion. Newspapers, magazines, and publishers besieged Dr. Bevis with offers. Some wanted his story, others wanted the women's stories, and still others wanted pictures of the test-tube babies. One publication alone offered Bevis something in the neighborhood of $75,-000 if he would grant an interview and tell how it had happened and what sort of women were involved. He regarded such offers as crass attempts to bribe him into betraying the identities of his patients and rejected all of them. When he emerged from the hospital, he announced that he was so disgusted by the scientific and public reaction to his work, work that he had regarded as highly humanitarian, that he would do no further research in the area of in vitro fertilization or embryo transplant. He was, in short, abandoning a large portion of his life's work.

My article for the *New York Times Magazine* on the embryo work had been delivered but not yet published when the Bevis story broke. Inevitably, I had to rewrite the piece to accommodate the story. I quoted Dr. Steptoe, who refused to accept the Bevis claim and said he was astounded by the announcement. Some researchers I spoke with implied that, if the Leeds scientist had got negative results or defective offspring, he would have made no announcement of his work at all. Others were unwilling to accept the idea that, by proceeding in secrecy, Bevis had been intent only on protecting his patients and their unique offspring. Some felt Bevis had also been out to protect himself in case something went badly wrong.

I acknowledged that possibility but also, giving Bevis the benefit of the doubt, pointed out that, with prominent individuals making statements of a condemnatory if not incendiary nature, it was not difficult to understand such secrecy and self-protection. For example, James Watson had just warned a congressional subcommittee that with successful embryo transplant "all sorts of bad scenarios" would erupt—that in the wake of these operations "all hell would break loose, politically and morally, all over the world."

One thing was certain. In the wake of the Bevis controversy all hell broke loose somewhere in Darwin's brain. I don't know whether he knew Bevis, but it was clear that Darwin took very seriously and, it seemed, very personally the reaction to Bevis's work. Darwin declared that he could have nothing further to do with our project. He seemed overwhelmed by what had happened. He said that he had been dreaming to think that he could become involved in such a project, that secrets could not be kept and that he would be ruined, as he said Bevis had been. He hinted darkly that Bevis had been forced by some undesignated power to disclose what he had been up to or face consequences even more unpleasant than public and scientific condemnation. Shettles's name came up as well.

A short time later Darwin met in private with Max in New York and was again persuaded to stick with the project. Max later told me that it was Bevis's survival (at least no one had yet strung him up, fired him from his job, or, as far as we knew, blackballed him from his professional societies or social clubs) which had, more than anything else, calmed Darwin. It was at this point, I believe, that Max also sweetened the pot a bit—promising Darwin funding which would reach beyond the current project into research areas of his own personal interest. Darwin made the most of his trip to the States and went on what Max called "a buying spree," obtaining more supplies and equipment for his lab. Partly at my suggestion, Max invited Darwin to come to California, where we could all get together once more.

CHAPTER 15

I met with Max in Marin County in early August while Darwin was still on the East Coast. My conversation with Max was a memorable one. It began with a recap of the Bevis affair, then turned to a discussion of a book I had recently written in collaboration with a South African doctor. It was about something called "abdominal decompression," an innovation then (as now) in wide use in South Africa. It was a technique which, when applied during pregnancy, increased oxygenation to the developing fetus and seemed to have the effect of minimizing the chances of

87

mental retardation. There was also anecdotal evidence that, by the same token, it might optimize IQ. But mainly it was being used to overcome and avoid toxemias of pregnancy and other causes of spontaneous abortion. Welcome side effects were marked diminution of pain and length of labor.

Max was largely interested in the still-speculative effects the technique might have on intelligence. He recalled that Huxley, in *Brave New World,* had foreseen the "predestinators" of the future fixing the intelligence of babies-in-bottles by regulating the amount of oxygen that bubbled through their nutrient baths. Max said he had decided that the surrogate who would carry his clonal copy must have the benefit of this new technique. He wanted me to help him secure the needed apparatus to construct a decompression unit or "suit," as the South Africans called it.

I couldn't be sure until some time later that Max was really all that interested in decompression. It was typical of him to begin a conversation on a businesslike note and then, only after having warmed up, to move on to whatever it was he really wanted to discuss. The subject that evening, it turned out, was to be twins. Putting my book aside, he picked up another. It was called *The Curious World of Twins.* [16] He opened it to a section on cloning, in relation to which the authors introduced a number of "devastating" and "bizarre" speculations, all of which, they added, made real twins "seem somewhat cozier and more appealing." I groaned but read another passage Max had marked. It noted that the word "twin" had an Anglo-Saxon origin and meant "two strands twisted together." Leaving clones aside for the time being, the authors asserted that even among ordinary identical twins "the tangle is not always untied, and the union between the two can be close, mysterious and at times unfathomable."

I observed that the word "twisted" certainly seemed to be often applied to twins and wondered if it could in any real sense be justified.

Max said that he had found the book to be well documented. He pointed to a case that had been clinically detailed in the

88

Archives of General Psychiatry. [17] I read the pertinent passages. This was the case of two female identicals, the weaker of whom copied the stronger in all things.

Both girls appeared well adjusted otherwise and did well in school. Later they both worked for the same company. Suddenly, the more aggressive of the two suffered an apparent nervous breakdown. She wept, stared vacantly into space, and was often oblivious of all that was happening around her. The other sister continued with her work at first, but soon began to display the same symptoms of mental illness as her sister. Eventually both had to be institutionalized. They were sent to the same place and even shared the same room.

The twins continued to disintegrate over a period of years until doctors, in a final effort to try to save at least one of them, decided to separate the two for the first time. The sister who had been the stronger was found dead, lying curled up in a fetal position next to her bed, on the first night of the separation. A few minutes later the other sister was found lying in the same position in her room—also dead. Extensive investigations failed to find any evidence of murder, suicide, or accidental poisoning. "Death," the authors concluded, "had seemingly come from within them." One of the psychiatrists said that he believed the sisters were "like two people with one brain." Max showed me other evidence of that mysterious medical event that he had mentioned to me, as I now recalled, at our first meeting nearly a year ago—the simultaneous-death syndrome.

My only response was to shudder. I knew this was leading to something, and we soon arrived there. Max said that when he was very young he had persistent dreams in which his phantom playmate looked exactly like himself. When he was asleep, he said, his double was always there. When he was awake he felt as though part of himself were missing. Later, when he learned that he had been adopted, he became convinced that he had an identical twin brother somewhere.

Over the years, with difficulty, he was able to trace his origins back through a series of foster families to a small orphanage and, finally, greasing the cogs of bureaucracy with cash, to a hospital and a courthouse where, he said, he found records that he believes mark his birth. The records indicated the presence of an identical twin. The father was marked "Unknown." The mother's last name had been deleted. And he was unable to find in the records of the orphanage to which he had been sent any evidence of another baby of his birth date and description. Nor could he find any likely candidates in the records of other orphanages in the vicinity. But this twin, whether real or imagined, was to haunt him for some time—until a dramatic event occurred.

At age eight or nine, he recalled, he had stopped having the dreams. The twin was gone, though Max continued often to think about him during his waking hours. There were no more dreams until one night when he was fourteen. The twin was suddenly back, looking precisely as Max then looked.

He could still recall the dream vividly. He saw himself walking through a field on the farm where he was living, and on the opposite side, walking toward him, was another boy—his twin. The field had just been furrowed and was wet with rain; he could feel the muck clinging to his boots. His steps became heavier and heavier as he approached his twin. He felt happy and sad and frustrated all at once, he said. He wanted to move faster. He wondered where his twin had been all this time, why he had been left so long alone.

When he was within several yards of his dream brother, he was stopped cold, horrified. The twin had a yellow cast to his skin and was coughing. Blood was just beginning to spill, slowly, from his mouth. Max said that he felt, in that moment, as if he were being ripped apart. No words were spoken. As the twin sank into the mud, Max could feel himself being dragged down. The twin was reaching out to him. Max wanted to turn and run, but he couldn't. In his dream he felt his own arms reaching back, but he couldn't quite touch the other. He was convinced now, he said, that if he had touched his twin, he would have died on the spot

90

—died not merely in the dream but in reality as well. As it was, he woke up at that moment screaming, drenched in sweat. The whole house woke up with him.

I asked him if he had ever had nightmares like that before. He said he hadn't, nor had he had any like that since. But there was more to come. He had been unable to sleep after the dream and lay in bed shaking with what, by morning, appeared to be chills and fever. His foster family was poor, and there was no doctor for a hundred miles. By the following night he was so weak he could hardly stand. Everyone was commenting on how yellow he looked, and he had started coughing. He didn't tell any of the family about his dream. Between fits of coughing he'd catch himself drifting toward sleep. That terrified him so much, he said, that he pinched himself until he bled—all through the night; he was not about to go to sleep and die, choking on his own blood, which he was convinced would be his fate if he did permit himself to sleep.

The next morning his foster parents looked at his arms and chest and saw dozens of "little bloody parentheses" where he'd pinched himself all night. His foster mother, whom he characterized as "a real Bible-Belter," called these gouges "the Devil's pox" and ordered the other children to stay away from him. Max had always been regarded by the others in the family as "unnatural" to begin with, he said, because of his supposed bastard origins. They were always expecting strange behavior from him.

A preacher was called in to exorcise the devil in Max. The man was big, raw-boned, and dressed all in black. He chanted, shouted, spat obscenities, finally fell down on the floor, and, Max said, all but frothed at the mouth. Apparently he was wrestling with the demon in Max's fourteen-year-old body. The upshot of all this was that Max burst out laughing. The preacher declared him "freed."

Within a couple of days Max was up and around and privately decided it was time for him to take off on his own. He was convinced his twin was dead—had died the night of the dream.

(He still believed this.) After that, it seemed to him, he no longer needed his foster family, needed no such make-believe link to his past, to his identity, because the only real link was now dead and gone. He was indeed "free." And this, he said, both thrilled and terrified him—and continued to do so to the present day.

In a sense, I pointed out, and he had to admit I was right, the search for the lost twin continued, too.

CHAPTER 16

From Marin Max went to his small ranch in southern California. Darwin planned to join him there briefly before leaving the country. I also wanted to be on hand and would follow Max down in a few days. Both Max and I were mindful, by this time, of Darwin's volatile character, and a meeting in which, as Max put it, we could all "press the flesh" and exchange views might be politic if nothing else. Personally, I was eager to learn firsthand how things were shaping up at "the facility," as we often referred to Darwin's outpost.

Except for the grounds right around the house, the ranch was a dusty spread, but it abutted on some state-owned mountain land and afforded the opportunity for ample hiking and horseback riding for those so inclined. Max was fond of both, and when I arrived (a car, again, was sent to the airport to pick me up), the Oriental houseman informed me that both Max and the doctor, who had also just arrived, were out riding. The house itself was a low, rambling structure with Spanish-tiled roof. An open, interior court contained an elaborate outdoor fireplace/barbecue pit and a small swimming pool. Several sliding glass doors opened from this court into bedrooms and living quarters.

After I'd unpacked I wandered over to a complex of barns

and corrals and soon discovered that Max kept not only horses but goats and chickens as well. An employee who was cleaning out the stables explained that Max got all his milk and eggs from these animals, even when he was in Marin County, because they were free of stimulants, antibiotics, and the other chemical additives found in the commercial stuff.

It wasn't long before the two horsemen rode up, leaving a trail of dust behind them, and I was treated to the sight of Max in cowboy boots and Stetson, which—as soon as my eyes adjusted to the sight—seemed to suit him well. Darwin looked a bit outlandish in slacks and a pinstriped shirt—obviously the most casual things he had along—and his boots, too big, were Max's. He was sweating profusely. But he rode well, remembering how, he said, from his youth.

We spent three days at the ranch, mostly sitting around the pool, where Darwin could smoke with less guilt. It was a chance for all of us to relax with one another for the first time—and to some extent get to know each other in a more casual way. Darwin, as usual, was concerned about what "they"—his peers, the press, etcetera—were saying about genetic engineering and cloning in particular. Various articles that had been appearing on these subjects served as initial conversation pieces. I cannot claim that we agonized endlessly over the responsibilities involved in our decision to go ahead with our project, but our being still so sensitive to outside comment reflected our continuing concern.

We discussed an article that had appeared earlier in the year in the journal *Nature,* written by noted molecular biologist Gunther Stent.[18] It was craftily argued in this piece that to oppose the cloning of humans was "to betray the Western dream of the City of God." In reality, I felt, the piece was not an argument *for* cloning but a sly critique of some inconsistencies the author perceived in Christian belief. However, if Darwin chose to be cheered by the article, neither Max nor I was very eager to dampen his spirits.

As time went along, we found it was often possible to predict

93

Darwin's mood by the tone of the latest article on cloning. If that tone was hostile, Darwin was likely to either go on the rampage or become sullen and depressed. The most memorable example of this was evoked by a piece that appeared in *Harper's* magazine in early 1975. In this article journalist Horace Judson lashed out at Bevis, Edwards, and others, accusing them of seeking fame while ignoring the rights of the fetuses upon which they experimented.[19]

Both Max and I, later on, clipped and sent to Darwin an editorial from a recent issue of *Science*. We laughed when Darwin announced that he, too, had seen it and clipped it. It was predicted in this editorial that mammalian cloning would be a reality fairly soon, but even human cloning, the editorialist said, would be "hardly terrifying." If human cloning were eventually prohibited, "an occasional violation would not shake the heavens."[20] Darwin took that almost as a commendation.

Fortunately, there were some articles that he apparently never saw, such as the one that would appear a few months after this meeting in the *New England Journal of Medicine* by Lewis Thomas, director of the Sloan-Kettering Cancer Center in New York. This piece, I felt, had to be considered something of a setback for those who were willing to graduate cloning from unalloyed monster to uneasy metaphor. "Selfness," Thomas wrote, "is an essential fact of life. The thought of human non-selfness, precise sameness, is terrifying when you think about it." And to those of his colleagues who might be thinking about it— that is, might be thinking about actually cloning a human— Thomas had a word of advice: "Don't." He added, "Fiddle around, if you must fiddle, but never with ways to keep things the same. Heaven, somewhere ahead, has got to be a change."[21]

Such articles, if they did nothing else, helped us to formulate our own views. It was in response to some of the negative comment that during this visit Darwin really opened up to me and spoke at some length and with some passion, obviously intent upon asserting the rightness of what he was doing. Max was busy

94

with a series of phone calls related to some of his other business, and Darwin and I were alone at poolside.

I asked him about the chances of creating a monster through the cloning process. If anything went grossly wrong, he felt, there would most likely be a spontaneous abortion—nature's way of eliminating defects. As a backup, amniocentesis and other diagnostic procedures would be used to ensure, or help ensure, that a healthy baby was gestating in the surrogate womb. He would not hesitate to perform an abortion if one seemed indicated. To him, obviously, the possible benefits outweighed the possible risks.

He spoke in detail of what he hoped to accomplish along the way, in terms of his own research specialty, an area that has important potential for human health. He said he hoped also that success might destroy some taboos that inhibited research with human genetic material in the study of such things as cancer. He hinted that he might eventually disclose his participation in the project, when and if the climate seemed right.

But, I persisted, couldn't cloning be regarded as unnatural? A baby born deformed or mentally abnormal by virtue of achieving life through this man-made aperture would be a black mark on the human race. Why any more so, Darwin wanted to know, than the retarded or deformed babies born every day? That, I insisted, was nature's doing. Darwin asked if I had excluded the human being from nature. When abnormal babies resulted from human sexual intercourse, he said, it was humans' doing, not the doing of trees, birds, or bees. What we humans do would have to be considered every bit as natural as the activities of those other living things. If we were to believe that we were evolving, generally making progress, then it could be argued that cloning was as natural as sexual reproduction—merely a new creative impulse we were only now becoming aware of and capable of fulfilling.

Darwin was getting excited. So many had claimed we were nature's crowning achievement, he continued, something set apart from and above the rest of nature's creatures by virtue of

95

our ability to think and exercise our will. If so, then cloning was something that might very naturally be expected of us and something that was a considerable improvement over sexual reproduction, in that it took a far greater application of mind and will to clone than it did to engage in sex.

Nor would he concede that cloning was unnatural in even the commonest sense of the word. The world, he said, was full of parthenogenones, and most of them got here without any help from us. These parthenogenones, which aren't too far removed from clones, are the offspring of so-called "virgin births," living things sprung from a single parent. Parthenogenesis had been observed in many species and had even been artificially induced in mammals as early as the 1930s by Dr. Gregory Pincus, later to become one of the "fathers" of the birth-control pill.

It was even believed, Darwin went on, that there were a certain number of human parthenogenones. Dr. Helen Spurway, a lecturer in eugenics and biometry (biological statistics) at University College, London, had asserted that one in every one to two million women were probably the product of true virgin birth. Dr. Spurway had laid down a number of requirements a mother would have to fulfill to prove that her daughter was of what the Bible had called "immaculate conception." (Jesus, by the way, Darwin pointed out, could not have been a parthenogenone. *All* parthenogenones arise from unfertilized female eggs and must, therefore, themselves be female.) Among other things, her blood group would have to be compatible with that of her daughter, and she would have to be able to accept skin grafts from the daughter without rejection.

Darwin pointed out the malleability and adaptability of living matter, the facts of which, he said, flew in the faces of those who had very firm ideas about what was natural and what was unnatural. People's views on reproduction, for example, were typically rigid: according to the common wisdom, there was only one true, moral, natural mode of reproduction; any deviation from reproductive heterosexual intercourse—"preferably performed,"

96

he added, in "the missionary position"—was damned as unnatural or immoral.

He told me of a report reliably on record that an apparent chicken cock had been burned at the stake in Basel, Switzerland, in the fifteenth century for having had the audacity to lay an egg. A good thing, he said, that people hadn't known in those days—and possibly in this day—about the slipper limpet, an opportunistic sea creature that can change sex at will. If there is an excess of males in its vicinity it can switch to a female, and vice versa.

The line between even the human sexes, he said, was not so cut-and-dried as many of us liked to believe. The human male starts out life as something of a hermaphrodite. At a number of points in his prenatal development the tiny male, his maleness protected only by the smallest of all chromosomes, the Y sex chromosome, is in constant danger of becoming a female. The penis and scrotum may for some time still "degenerate" into the more basic labia and clitoris of the female.

"Fetal malleability," as Darwin called it, was so pronounced that scientists had been able through simple hormone injections to induce dramatic sexual transformations in a number of mammals. Genetic males were made to change sex entirely, being born with functioning ovaries and vaginas. In some other animal experiments, "false females" which were in reality genetic males, prenatally transformed by hormone intervention, had been coupled with true males, and false males with true females. From these couplings were born offspring which scientists said were the issue of *two* fathers or *two* mothers. Jean Rostand had referred to these as "homosexual unions" and had observed that they resulted in only male offspring when two males mated and in only female offspring when two females mated.

All this, Darwin concluded, showed that nature was not nearly so dogmatic as some of its smartest offspring, namely humans, when it came to deciding what was natural. Considering both the exotic, unaided variations of sexual and reproductive expression in nature and how easily nature was manipulated when

97

a scientist wanted to create other variations, Darwin was convinced that cloning, for example, could not be categorically characterized as perverse or improper, and those who would do so lived in ignorance of what he called "the real facts of life."

During this visit Max opened up some, too. His views were seldom so cogently expressed as Darwin's, and his position was by definition self-serving. He recognized this and I believe found it embarrassing to take a formal stance, fearing that he might sound hypocritical. He was obviously interested in justifying himself, but he usually approached this through a more distant treatment of the issues. His views were both like and unlike Darwin's.

The doctor would argue that we humans, given our ability to think and exercise our will, have a mandate from nature to participate in and even guide our own evolution, genetic as well as social and economic. At heart this is nothing other than the humanist ethic that would salvage mankind from a savage, dog-eat-dog world in which the fittest survive. Max, while recognizing the same mandate, would insist, however, that the salvage operation had become subverted by overzealous do-gooders whose actions foist upon us "gradual destruction by the unfittest." It was Max's view, shared by some sober geneticists, that the gene pool is being polluted by defectives nature might previously have weeded out. Max was fond of quoting scientists who foresaw a genetic twilight precipitated by our ever-increasing ability to keep alive the fertile possessors of previously fatal or debilitating genetic disorders. There were signs, some warned, that these detrimental traits were being spread throughout the population.

Max clearly saw himself as a rugged individualist, and so he could not be entirely comfortable with the concept of positive eugenics, embracing those measures by which we might improve, through selective breeding and the like, our genetic lot. Eugenics, at base, is a social program; it asks individuals to set aside many of their particular desires for the good of society—and thus is somewhat at variance with the impulse and destiny of the individual.

Max, after all, was one of the fittest who had survived—a sort of crypto-Social Darwinist, an adherent of the philosophy espoused by Herbert Spencer in which success in all things, including finances, is regarded as a consequence of superior heredity. But Max, I guessed, had enlarged and mystified the concept to embrace superior will, as well, since he conceived of himself as largely a creature of his own imagination.

I found that Max was able to advocate programs of genetic intervention only because those programs did not touch him personally. He regarded himself, I came to believe, as a model—not as one of those to be molded by eugenics. In his vision of participatory evolution, the fittest would participate first. This was not his most attractive aspect.

During our visit to the ranch, Darwin seemed reluctant to say much about his research. It was too early, he said, to make any estimates of possible or probable progress. He was happy with the facility but noted that more equipment was needed. All of his needs, Max assured him, would be taken care of at once. Darwin said he enjoyed the tropical climate and was adapting rapidly to his new locale. The living quarters provided for him at the facility were quite adequate.

The research he had already begun was aimed at finding optimal growth cultures for eggs. He was also involved already with enucleation experiments—work directed toward the removal of nuclei from eggs in preparation for the implantation of body cells. I asked him which animals he was working with. This question was to come up again later, and I recall now that on this first occasion he was very careful in phrasing his answer. He said that he was in the process of "building up" colonies of mice and rabbits. I assumed at the time that this meant he was experimenting with the eggs of these animals. When I asked him at which point he would begin working with humans—a matter of some concern to me, since I felt certain risks must first be ruled out in the course of doing animal work—he waved his hands and shook his head as if I were jumping too far ahead. Again, I assumed, he was saying it was too early to consider this matter. His immediate objective would be to get an enucleated egg ready to accept a body cell and show some sign—any sign—of being willing to divide.

He had also been busy procuring help. With Max's permission he had already taken into his confidence a female doctor who would soon be working with him full time. He had also just arranged to obtain the assistance of a man who did not want his identity known—even to Max. This man, whom Darwin described as "one of the top men in his field," had worked with Darwin on previous occasions and was, in the doctor's words, "absolutely trustworthy." In any event, he would not be told who Max was, either; in fact, Darwin claimed, he wanted it that way. This man, whom I will hereafter refer to at times as Darwin's "shadow colleague," would never actually visit the facility. He would collaborate with Darwin by phone, perhaps by letter, and occasionally in person, if necessary, at meeting places known only to Darwin and himself. The colleague, meanwhile, was promising to supply Darwin with one of *his* top assistants—a young man who had done graduate work in an area in which Darwin said he particularly needed some outside assistance. Whether Max's identity would be made known to *this* assistant had not yet been determined.

When I expressed some concern in private about these "new faces," as I described them, Max said that he wasn't worried. They had been selected very carefully, and Darwin considered them essential. He added that through his entire business career he had succeeded in keeping a very low profile in large part because he was good at precisely this: selecting his help carefully. On those rare occasions when he had found it necessary to show any significant part of his hand to other players, he had secured their loyalties by retaining them for life. No one, he said, had ever complained.

PART II: METHODS

*The "genetic engineer" about whom such fears
have been expressed in public will not be the
geneticist who has discovered the experimental
techniques that open up such staggering vistas.
The genetic engineer will be the man who applies
the new technologies. It will be, I assert, the
venturesome surgeon whose intrepid step opens
those new doors.*

—Dr. Bentley Glass

Each of us is someone's monster.

—Paul Chauchard

CHAPTER 17

The subject of participatory evolution, which saw man with his hands down his genes "fiddling" with himself, to use Lewis Thomas's evocative verb, was one that occupied a great deal of my time. This brave new world we might be helping to create was going to be a strange place, and I wanted to make sure we were all aware of that. My task, before I became more deeply involved in interpreting for Max some of the methodologies Darwin would explore, was to put into some sort of context our part and our place in this biological revolution that R. C. W. Ettinger *(Man into Superman)*[1] had written would effect "a sharp discontinuity in history."

There were developments galore that already made me squirm. Many of them, I could tell, troubled Darwin, too. The more uncomfortable I could make all of us, I thought, the better. The more we squirmed and the more we tried to disassociate ourselves from some of the other, often astonishing goings-on in the biological world, the more mindful I hoped we would be of the far-reaching implications of our own effort—and thus the more cautious and responsible.

Yet sometimes it seemed that, on the contrary, the memos and reports I shot off to Max and Darwin served only to give comfort. With so many seemingly incredible things having already taken place or been seriously proposed, we could easily come to believe that what we were doing was not terribly unconventional or out of step, after all.

Since my reports were, at first anyway, mainly for Max's benefit, I started out by dealing with what I called "ancient history." Max already had some knowledge in this area. Predictably enough, one of his heroes had been Dr. Hermann Muller, winner of the 1946 Nobel Prize in physiology and medicine for his pioneer work in using X rays to cause mutations. Muller was later noted for a plan he named "germinal choice," which called for parents to give up their egotistical desires to reproduce their own genetic traits and opt instead to have children via artificial insemination. This would ensure the continuing evolution of man, Muller believed, since only the strongest and most desirable men would be permitted to contribute seed to the frozen-sperm banks he advocated.

Muller predicted that eggs, too, would one day be readily available and that embryos themselves could be frozen and stored, then thawed out and used as needed. (All these things have come to pass in animal work and could, no doubt, be extended to humans in short order.) In sharp contrast to current policies of strict sperm-donor anonymity, Muller called for detailed descriptions of all egg and sperm donors and proposed that these be made available to prospective parents so that they could make "wise" decisions. He envisioned catalogues parents-to-be could flip through in search of the ideal egg to be mated with the ideal sperm. These catalogues would contain not only descriptions of the donors but even pictures of them.[2]

I told Max that since Muller's day a number of others, most notably Dr. E. S. E. Hafez, an internationally known experimental reproductive biologist, had moved a long way in the direction of realizing those frozen-embryo banks. His successes had given him enough confidence to predict that prospective parents could soon, if society permitted, select babies from day-old frozen embryos guaranteed free of all genetic defects and described as to sex, eye color, probable IQ, and so on. All this—and pictures of what the grown-up product could be expected to look like—would be on labels affixed to the packages that contained the frozen person-to-be. Following purchase, the embryo would be thawed and implanted under a doctor's care.

In the more distant future, the economical Dr. Hafez suggested, we could realize great savings in our space-exploration efforts if we would include in the cargo, on particularly long journeys to distant planets we hoped to explore and possibly colonize, a group of frozen human embryos. No need then for a lot of cumbersome and costly life-support systems, and no need to worry about the future colonists getting bored on that long journey. They could be thawed out, grown in test tubes, reared, and instructed as to their missions in life by computer surrogates upon arrival.[3]

I then mentioned to Max that population imbalances, both too many babies and too few, might contribute powerfully to the development of programmed births employing selected eggs, sperm, or even frozen embryos. The market in human sex cells already existed. Medical students and others had been selling sperm for years. Eggs, now that they were becoming available, would similarly be sold. And I had learned, in the course of researching a magazine article, that there were black-market baby brokers who showed their clients what amounted to breeding catalogues. For a fee, often as high as $25,000, one could select from a group of attractive young men and women both the father *and* the mother of the child one would adopt ("buy" was a better word, I felt) on the day it was born. What with the current difficulty of obtaining healthy white babies for adoption through normal channels, clients were not in short supply.[4]

It had also been suggested that the day would come when famous individuals would sell their eggs and sperm the same way they now sold their names to endorse products. I had no doubt that there would be buyers and that the price for the eggs and sperm of movie stars, noted athletes, acknowledged geniuses, famous dancers, eminent artists, Olympic medalists, and the like would be high indeed. Even fertile couples could be expected to get in on the bidding. I thought this would come to pass whether society officially approved or not.

On the other hand, I told Max, the more pressing problem of too many babies was encouraging an increasing number of well-credentialed individuals to advocate plans that would prohibit unlicensed reproduction. The British Nobel Prize winner Francis Crick and the noted American economist Kenneth Boulding, among many others, had spoken favorably of such plans. One psychologist, Roger W. McIntire, writing in the journal *BioScience*, had proposed a plan whereby licenses to have children would be granted on the basis of contests that would test skills (and thus overall "worthiness" to reproduce) in such areas as math, the arts, sports, and business.[5] A similar plan was put forward at the American Psychological Association in 1974, including detailed proposals describing a special federal regulatory agency that would be charged with administering some of the long-term contraceptives now under development to all females (and possibly all males) of reproductive age. Only those licensed to have children would be given the necessary antidote to the contraceptive. This was known as the "Lock-Unlock Scenario." I was confident it was tucked away in a number of government files and think tanks against the day when implementing it might be "necessary"—or politically possible.

I mentioned Dr. Jean Rostand earlier. He predicted "telegenesis," the creation of life achieved by the union of two sex cells obtained from points distant in space, and "paleogenesis," birth from germ cells obtained at different points in time, or well before conception. These things, too, had already come to pass. For instance, there was the press story of a woman who had given birth to her late husband's child nearly two years after he had died of cancer. He had thought to leave some of his sperm stored in liquid nitrogen.

Now Dr. Bentley Glass, observing that such had already been achieved using other mammalian tissues, predicted that it would soon be possible to culture in the laboratory portions of human ovaries and testes sufficient to provide continuous production of egg and sperm cells. Such a development could only facilitate the market in germinal material and prefabricated babies.

According to another report, the genetics experts were making such strides in the realm of gene-splicing and hybridization that some had felt it necessary to warn that human material could some day be combined with other animal material to create human-animal "chimeras." "Before long," Dr. Lederberg had stated, "we are bound to hear of tests of the effect of dosage of the human twenty-first chromosome, for example, on the development of the mouse or gorilla." In this fashion, he feared, we would gradually produce chimeras of "varying proportions of human, subhuman and hybrid tissue."[6]

Meanwhile, the ever-pragmatic Rand Corporation had predicted that "parahumans" would be created in the lab to perform low-grade labor. A report on this prediction observed that some employers were already hiring the mentally retarded because they were the only people willing to stick with—and even take some pride in—such menial tasks as floor mopping.

Even more alarming, I told Max, were reports on recombinant DNA work (that research which Nobelist James Watson was later to defend so testily, as I mentioned in chapter 7). One report I described to Max suggested the possibility of using recombinant DNA procedures to create humanoid creatures which would be used only to provide a ready supply of organs for transplanting and medical experimentation. The idea seemed to be that such creatures, because they might legally be defined as nonhuman, could be sacrificed without anyone's being charged with murder.

Reports like that made all of us cringe. But there was more. The list of "marvels" or "monsters" that were emerging from the lab was a long one. It wasn't simply the human body but also the mind and, some might argue, the soul that were up for remodeling.

Neurosurgeons at Cleveland General Hospital had succeeded in transplanting the head of one monkey to the decapitated body of a second monkey (which was still living, thanks to artificial life-support systems). This newly constructed monkey lived for a week, until the head "recognized" the body as "foreign" and rejected it (or vice versa) in the same way that transplanted hearts were often rejected. Once cloning was established, transplants among members of the same clone, including even head transplants, would be free of the rejection syndrome. And a famous surgeon, I continued, had stated that human head transplants would be possible in the very near future. (The technical details, he commented later, could be worked out in "a year, no more."7)

There were some very good reasons to be "fiddling" around with things such as head transplants (we could learn a great deal about how the central nervous system works, for example), but you had to worry when you learned (from a Rand Corporation report) that the Soviets were experimenting with disembodied cat brains, linking feline gray matter to artificial maintenance and sensing systems, and creating biocybernetic guidance packages for implantation in air-to-air missiles! The cybernized cat brains, the Soviets seemed to hope, would be able to recognize optical impulses emanating from their targets and to transmit guidance signals accordingly, so that the missiles always stayed right on target.

How much longer before human brains might be used for the same sort of thing? Scientists had made considerable progress in keeping disembodied monkey brains alive for long periods in the laboratory. You couldn't even begin to guess at the thoughts of an isolated or computer-augmented human brain. The potential for horror seemed almost unlimited, yet experimentation was going forward in dozens of centers around the world that might one day lead to this new order of disembodied "man."

For example, work was under way at the Burden Neurological Institute in Briston, England, which had shown that human brains could be linked to computers via electrodes and a "rapport" established sufficient to enable the mind to stop, start, and otherwise operate machines merely by "thinking" commands. This could be extremely useful, no doubt, but it had also been demonstrated that machines, computers, and other electronic apparatus could equally instruct and control the human brain.

At centers all over the world experiments were proceeding with ESB, "electronic stimulation of the brain." Humans had already been surgically implanted with hundreds of deep-driven brain electrodes of the same sort that had made animals, in the words of Dr. Jose M. R. Delgado, then at Yale University School of Medicine, "perform a variety of responses with predictable reliability, as if they were electronic toys under human control."[8]

ESB could be useful for mapping the various areas and functions of the brain, but its potential for abuse, I told Max, seemed to me to be enormous. With it, you could at the push of a button evoke in another person great sexual desire, hunger, fear, hatred, or rage. You could even cause the subject to see visions, hallucinate, and "remember" things that in reality had never been experienced.

I outlined plans that had been contained in government monographs (at least one of which was concealed from the public); one plan, proposed by a Harvard psychologist who described himself as a "social gadgeteer," would literally bug the brains and bodies of individuals who were or merely threatened to be out of step with society. According to this plan, sensors were available that could detect even "inappropriate" erections—and electrodes, implanted in the brain and attached to the body, could be used to deliver "corrective" kilovolts of electricity whenever the subject responded to his environment "improperly."

Since I had criticized this and similar plans in two magazine articles I wrote on psychotechnological abuses, I felt that when I revealed my role in the cloning project some might accuse me of hypocrisy, of purveying a convenient double standard: it was all right to manipulate the genes and to clone, but it wasn't all right to manipulate behavior by pulsing electricity into the brain.[9]

But this was what it was all about, I thought; one had to look carefully at each of these awesome new technologies, weigh the risks and benefits of each, individually, and then decide which was good or bad or neutral in any given set of circumstances. I felt that I had exercised my best judgment. I knew that I *might* be wrong.

One new breakthrough that we looked at particularly carefully seemed to have resulted in a quasi form of mammalian cloning. The method used was "embryo fission." The principal researcher behind this startling work was Dr. Beatrice Mintz of the Institute for Cancer Research in Philadelphia. Her work was shedding important new light on the basic nature of cancer. She was also, incidentally, shedding new light on some unique ways of creating life; she had not only cloned by taking embryos apart but had also created animal babies with not two but *four* parents —by fusing embryos together!

Using the eggs of mice, which, though not quite so small as human eggs, are nonetheless very small and difficult to work with indeed, Dr. Mintz had accomplished some unusual things. In her fission work she had, for example, taken 2-cell mouse embryos and microsurgically disassembled them under a dissecting microscope. She had taken the single cells that were the product of this disassembly and successfully implanted each of them into the wombs of surrogate mothers. Thus, what had been destined to be but a single mouse became, instead, *two* genetically identical mice. She was able to do this with even 4- and 8-cell embryos, as she refined the technique, creating identical quadruplets and octets. More highly developed embryos had so far resisted disassembly and regeneration by this method, but she had shown that you could still create hundreds of identicals by persistently taking apart 2- to 8-celled embryos, letting them grow, and then disassembling them again—on to as many as you wanted.

Again in the process of creating useful biological test-beds for her study of cellular differentiation and malignancy, Mintz had also induced the *fusion* of embryonic cells, in such a way that there resulted mice which had not two but four genetic parents. She took an embryo created by the normal mating of two pure-bred white albino mice and placed it in a test tube with an embryo which was the product of two purebred black mice, also mated in the normal fashion. Then into the test tube containing these two embryos she poured some chemicals that caused the "cement" holding together the individual cells of each to dissolve. Next, the disconnected cells were mixed together and, after a time, began to join forces. Oblivious of their original liaisons, they ultimately formed one unified embryo.

Next Mintz took this "communal zygote," as I called it, and implanted it into a mother-mouse surrogate. The embryo took hold, developed normally, and emerged as a healthy baby mouse which also grew and functioned normally. It did, however, look rather odd. Unlike its two sets of purebred parents, it had an almost striped appearance, with the light and dark contrasts evident even in its eyes. Mintz and co-workers subsequently created hundreds of these polyparented creatures.[10]

Darwin agreed that if this could be done in mice it could almost certainly be done in humans. The day might come, I said, when children would literally be "constructed by committee," perhaps under the guidance of a computer analysis indicating which individual genetic traits in aggregate would create a person of the currently desired qualities. Dr. Mintz herself said she was quite sure that all of this animal work could be duplicated in humans, though she didn't think it *ought* to be.

Some positive eugenicists, however, were bound to think embryo fusion quite a keen idea, and I had no doubt that, like cloning, it would become a human reality in the not-distant future. I hoped that the first to try it would be mindful of Dr. Mintz's warning that great pains would have to be taken to match pigments, sex chromosomes, and the like, or the resulting offspring might emerge hermaphroditic and, if not actually striped like a zebra, at least weirdly mottled in skin, eye, and hair appearance.

It was clear that we were not only entering but were already through the door of a new world in which man would become, "under the magic wand of biology"—as Jean Rostand had predicted so many years before—that "strange biped that will combine the properties of self-reproduction without males, like the green fly; of fertilizing his female at long distance, like the nautiloid mollusks; of changing sex, like the xiphophores; of growing from cuttings, like the earthworm; of replacing his missing parts, like the newt; of developing outside his mother's body, like the kangaroo; and of hibernating, like the hedgehog."[11]

These visions, in fact, seemed already quaintly dated and benign. They said nothing of humans that might add or subtract another's hopes, thoughts, and dreams at will, delete or compound genes to create greater or lesser beings, hybridize with other creatures, and meld both creatively and coercively with machines. I could only hope that what we had set about would come to signify, in however large or small a moment history might ultimately accord us, the mindful, and not the mindless, application of an awesome new force at work in the universe.

CHAPTER 18

I first visited the facility in December of 1974. The locale was very beautiful—a lush, tropical garden. As I was chauffeured from the airstrip to the lab and hospital, I found the jungle interrupted here and there by large expanses of rubber trees and coconut palms. On steep hillsides were trees I had never seen before but which my driver, a fellow named Roberto—who turned out to be a longtime employee of Max's—identified as nutmeg. Occasionally the narrow, paved road would take us within sight—and smell— of the ocean. That disagreeable odor, Roberto said, was fish drying in the sun.

I asked Roberto why the hospital seemed to be away from the major population area near the airport. It seemed that we were heading increasingly into the bush country. He said that Max and the others who had set the hospital up ten years earlier had asked the local people where they wanted it, and then had complied with their wishes. They wanted it close to their fishing grounds and rice paddies. Even many of those who worked in the town near the airport, Roberto added, preferred this "country" hospital to the one run by a religious order in the town (which, in population at least, was closer to being a city).

The location that was finally designated suited Max fine, Roberto said. Many of these rubber plantations were his. He even had an interest in the local fishing industry. Most of those who worked on the plantations did not really know whom they were working for—but they were happy that "the company" had provided them with a hospital. It made them more productive. Max was a very clever man, Roberto said. His biggest local interests were the factories, where, among other things, some rather complex equipment was assembled. "Cheap labor," Roberto said. Now other corporations were moving in to take advantage of it, too. Max, Roberto said pridefully, was always the leader.

My efforts to get Roberto to divulge specific information about Max's businesses, however, proved fruitless. He was obviously a very loyal employee. When I asked him whether Max was the sole owner of the hospital, he said that it was run by a foundation. Max's foundation? "Just the foundation," Roberto said, shrugging. And then he showed me another nutmeg tree.

There was one thing I should know, Roberto said as we neared the facility. At the hospital, which Max was visiting more often now, it was required that one always refer to Max simply as Mr. Smith. He was known by this name to various of the employees, who had always been told that he was a representative of the foundation that endowed the facility. Max, Roberto said, had his "private ways."

I would be less than candid if I claimed that my first impression of the facility was a good one. A grass hut, I think, might have pleased me more. It was constructed of concrete blocks and had a roof made of corrugated metal sheeting. It was literally an eyesore, for the sun reflected painfully off the roof and the uneven yellow paint of the walls. The central structure had two stories. Attached to it was a sprawling single-story wing. The duplicate of this was now under construction on the opposite side, though no workers were present when I arrived.

The parking lot in front of the facility was paved but pocked with chuckholes. Just off the asphalt where the jungle closed in on the small clearing were the rusting, rotting remains of an old truck. Deeper into the bush were some huts where people apparently lived. Just across the road was the edge of a rubber plantation.

The place seemed almost abandoned—"because of Christmas," Roberto explained. Christmas was only a couple of days away. I said that in a non-Christian country I had expected business to be as usual. It was still a holiday, Roberto said. I wasn't entirely surprised that Darwin didn't seem overjoyed to see me. His coolness was in line with the way obstacles had kept cropping up to my visiting earlier—some of them placed in my path, I was sure, by Darwin or Max or both. My insistence upon an on-site inspection had intensified as reports from Darwin's outpost became increasingly vague with each passing month.

My complaints at first evoked Max's sympathy, and I was led to believe that he, too, was a bit exasperated with Darwin's increasingly independent ways. But gradually Max's commiserative attitude gave way to one that seemed to me almost defensive. He had promised Darwin a free hand, he said, and could not always be nagging him for a "blow-by-blow" account of his progress. I said I would be satisfied with a "round-by-round" summary. But as much as several months into the project, for example, I had yet to learn what sort of animal experiments were under way or planned. We had all agreed that these would precede any actual human work.

When my requests to visit the facility were repeatedly set aside, Max would always assure me there was no problem. It was just that Darwin would be out of the area on the date I specified —buying new equipment or conferring with a colleague. Or if I would only wait a few weeks I could go with Max himself. Or there would be some sudden setback that made my trip "inadvisable."

My suspicions were mounting, particularly since I knew that Max had made several trips to the facility during this period. His excuses for not asking me along (had to leave unexpectedly, was conducting other confidential business on the same trip, and so on) were never thoroughly convincing. Through all of this he insisted that I was missing absolutely nothing. Everything seemed perpetually to be in the gearing-up stage. The closest I ever came to getting anything concrete on the animals was news that a large colony of mice Darwin was maintaining had been wiped out by some mystery disease. This was cited on one occasion as another reason for me to postpone my oft-projected trip.

Then suddenly, in the late fall of 1974, all resistance to my visit abated. My annoyance at Max and Darwin for now assuming an attitude of bafflement over my not having visited sooner was overcome only by my relief that whatever problems had stood in the way of my trip were apparently now solved. It did not occur to me that, in suggesting that I stop off in another country en route to the facility, Max might have had an ulterior motive. I had always wanted to visit that country, and the fact that Max had a residence there, at which he said I could make myself at home, made the suggestion irresistible. The effect of this stopover, however, was to delay my visit with Darwin still further—until the holiday.

Darwin lamented or pretended to lament that .nothing

would be happening for several days, as his staff was taking time off; he had thought he'd get them to work through the holiday or at any rate take not more than a day or two but . . . he shrugged it off as one of the vagaries of working in what he called "an uncivilized land." I began to sense that I'd been had.

Though my irritation was mounting, I made every effort to be civil. But Darwin, even edgier than I was, seemed to take offense at practically everything I said. When I commented on the seeming modesty of the facility he acted as if I had insulted him personally. He began quoting me the prices of various pieces of equipment, many of which, he said, had been engineered to his precise specifications. He lacked for nothing, he insisted, that might be found in the world's leading fertility-research clinics, and in the adjoining hospital it was now possible to perform the most sophisticated gynecological surgery. He had "dozens" of beds at his disposal, a full nursing staff and several highly skilled —and, he said, highly paid—technicians and assistants, some of whom had been recruited from afar. He had only to ask and he received. I began to sense that for Darwin—indeed, for any researcher who had ever been caught up in the maddening, paper-shuffling contest for funds that were ever harder to come by—this must all seem a bit like magic. Not only could you have what you wanted, but you could have it in a hurry without playing scientific politics, unwinding bureaucratic red tape, or submitting to peer review.

It was nice, Darwin acknowledged, to be independent. The lab even operated under the power of its own electrical generators, he beamed. He flicked a button on what appeared to be a small TV; instantly I saw sperm cells, greatly magnified, whipping across the screen. This, Darwin indicated, was a live performance. He nodded toward a nearby microscope setup with attached camera. As we made the rounds he all but caressed the recently acquired equipment: centrifuges, incubation chambers, cryogenic cooling equipment, laparoscopic fiber-optic viewing devices, row upon row of physiological and chemical substances—some of them marked radioactive—an electron microscope, a spectrophotometer, X-ray and ultrasonic equipment, an elaborate microsurgical theater, even a biocontainment chamber for experimentation with possibly infectious viruses.

The physical plant itself could use some sprucing up, Darwin conceded, but it was what the facility contained that counted. Indeed, to the untrained eye the work space in which Watson and Crick had climbed the spiral staircase that was to be the double helix of DNA would also have appeared ordinary if not downright shabby.

As the minutes ticked awkwardly by, it began to seem to me that, holiday or no holiday, very little was going on at this facility. Darwin seemed almost beside himself trying to find things to show me. I wondered when he was going to take me to see the animals I assumed he was working with. For that matter, I was beginning to wonder where the animals were housed. I'd seen practically every corner of the lab, which was situated at the end of the wing that jutted out from the main hospital structure. Darwin had even shown me his own quarters—two rooms on the top floor. Again, he said it was too bad I had come when I had. Things wouldn't be picking up for several days.

Piqued, I said that I might settle in for some time and that the holiday hiatus was thus of little concern to me. This suggestion seemed to cause Darwin some discomfort, but nothing like the discomfort my inquiries about the animal work seemed to provoke. Each time I would turn to this subject, he would miraculously find something else to talk about. He began suggesting places where I might go to spend the holiday. Finally I insisted point-blank upon seeing the "animal wing" he had briefly mentioned earlier in our conversation. When he had been offended by my comment about the modesty of his outlay, he had included an animal wing in his list of all the things Max had provided. I said now that I wanted to see it.

I halfway expected him to change his story about the new construction that was going on and claim that *this* would be the animal wing, even though he had already said this was an extension of the hospital facility. Now that he had no choice but to put up or shut up, he began making deprecatory remarks about the advertised animal wing. It wasn't much, he had plans to expand it later, and so forth. All this struck me as most puzzling, particularly since he had earlier boasted that with the budget Max had provided him he could afford a whole colony of costly primates. It seemed to me that the animal wing would have been his first order of business.

I don't know which of us was the more irritated as we made our way down a dark corridor to the back of the building, where, in something little better than a lean-to, were several cages, some of which, granted, were large enough to house a great ape or two. But the only creatures sulking in the dismal ambience of this place were a few conspicuously plump rabbits which were, I was soon to learn, destined to be sacrificed not in the pursuit of science but in the cooking pot. Darwin muttered something about all the trouble he'd had keeping animals alive and that he'd order no more until new facilities were built.

At this, the sinking disappointment that had merely been creeping over me for some months settled in with a hard thud. It seemed that nothing had been accomplished in what was now the better part of a year of work. Some impressive equipment had been assembled, but there was no evidence that it had been put to any good use. Those who knew Darwin often characterized him as an organizational genius. Why then did this effort appear to be floundering?

Darwin is not the sort of fellow you take by the lapels of his laboratory smock and shake vigorously while demanding to know what the hell he's been doing all this time. There is a touchy and sometimes intimidating aspect to the man that I had learned to deal with cautiously. He could be jolly and outgoing one moment, bellicose or withdrawn the next. His pride and ego, I knew, were enormous. That was why he could not countenance my (to him at least) disparaging remarks about the modesty of the facility which he obviously identified as his baby. It was this pride that gave me the opening I needed, as it turned out. If he had everything he wanted, if Max was cooperating to the fullest with funding, why then the delay with the animal work? I phrased my words gingerly, but the message was that I perceived some shortcoming—without identifying its source.

Darwin took the bait, instantly defensive. Things weren't going *that* badly, he said. In fact, he added, he had managed some "minor miracles," he and his assistants. And they had done so in an amazingly short time. He had assembled an excellent staff, obtained the best equipment money could buy, established a remarkable liaison with a former colleague whom he again characterized as one of the top men in the field, and, in general, had "squeezed five years of research into one." He was running this operation, not Max, and he could, he declared, spend two, three, four million dollars, whatever it took. Max had taken the lid off; he wanted results *now*. Darwin's tone was almost angry. I asked what the problem was, then—with all that money at his disposal, I would have expected a whole colony of primates in various stages of experimentation.

Darwin's look was one of exasperation and scorn. He did not lack for *primates*, he said, bearing down on the word. He had all he wanted and, again, they were "the best money can buy." I waited expectantly.

"Homo sapiens," he said tersely.

I was instantly astounded by my naïveté. They had been working with humans all along. My emotions in that moment were a dizzying mixture of anger, fear, and elation. Anger because we had agreed on the animal work. Fear because I could imagine so many abuses. Elation because this meant that rather than flounder we might actually be ready to fly. At first, however, I was most conscious of the anger and fear.

Darwin was ahead of me and any possible recriminations. It had all been silly, he said, to keep this from me. He had told Max so "time and again." This I doubted. But Max, he added, and this

I bought, had felt that what I didn't know wouldn't hurt me. Work with monkeys and apes, even mice and rabbits, Darwin went on, would have been costly and time-consuming and would have availed little. He rattled off various studies showing, he said, that many results with even the other primates were not predictive of results with humans, anyway. The same risks would still be there after months or years of primate work; all possible precautions were being taken; he was proceeding in an orderly step-by-step fashion; he was making real progress—actually he was "close, very close."

And his "primates," I asked, were they being informed of what was going on? They knew that he was after their eggs, he answered, at least in most cases, or that he wanted to use their wombs. No one had suffered. Again he became agitated, defensive. He was not only doing research, he said with a hint of piety; no, he had fallen also into the role of chief gynecological surgeon for the whole area—he was practically a "country doctor," dammit, up in the middle of the night, taking on one emergency after another. Women were flocking to him for tube-tying operations. If he saw a chance to get some eggs in the course of carrying out some other procedure, he took it, naturally.

All this, I knew, was meant to disarm me with his humanity. The truth probably fell somewhat short of his claims, but I could easily believe he had taken on some extracurricular work, if only to help maintain his cover. He was supposed to be doing a study on fertility patterns and contraception in the area while training the hospital staff in some of the newer and more sophisticated diagnostic and surgical procedures. He was ostensibly "rotating" through the facility on a grant from a foundation Max controlled.

I asked about compensation for the women. Darwin threw up his hands impatiently. In one hour in his laboratory or operating room, he said, they were paid more than they made in a month working on one of the plantations, farms, or factories in the area. And as for those factories, he added, the risks to health from working in them and being exposed to the dangerous machinery and noxious chemicals some of them employed or generated far exceeded any risks encountered in his facility. "My women," he said proudly, "no longer work in those places."

I later confirmed that the women who took part in Darwin's "study" were being paid sums that were very generous by local standards. I also discovered that the women who were selected for Darwin's purposes were often the youngest and the prettiest of the factory and farm workers—with Roberto doing the selecting.

And what about the bench embryos—those experimental creations that were "jettisoned"? Yes, of course, there had been some, Darwin said, pacing about. Many? Several, yes. Dozens? Maybe more than dozens, he said, shrugging; who knew? I suspected he did. I suspected he could account for every egg he had obtained and every embryo he had created. He had created some, hadn't he? In a sense, yes, he said. What of his feelings about discarding them at the end of their usefulness?

He looked exasperated again. He was opposed to casual abortion, he said, but as far as he was concerned this was not abortion. An embryo a day or two old was not a person; it was not, for that matter, even a fetus. When I asked what the difference was, he gave me a textbook definition: most held that an embryo did not become a fetus until the second trimester of pregnancy. Darwin said he did not consider an embryo any more a living, viable creature than he did an egg or sperm cell—or a strand of DNA. Well, when did it become a viable creature? Once it was attached, he said; once it had a literal grip on life, its trophoblast cells digging into the wall of the womb to suck up blood and oxygen. Then he would have to account the embryo a purposeful thing, driven as much as he was, in its own way, to survive. Thus he would consider even most first-trimester abortions immoral, but not the "destruction," if I must insist upon that word, of a preimplanted embryo.

Well, I asked, had he got to the point where he had tried implanting an embryo, and, if so, didn't he worry about the fate of those embryos that did not survive? Yes, he had got to that point, he answered, but the reason none had survived was that none had become implanted, and thus, by his definition, none had ever really lived. I told him I wished I could be as confident about all this as he was. With some visible effort at forbearance, he reminded me that he had lived longer than I and experienced a great deal more, and that all his professional life had been spent observing the behavior of eggs, sperm, embryos, and fetuses.

I could have pressed the point beyond Darwin's semantic defenses, but to do so, I knew, might well cause a major upheaval, might possibly even (though perhaps here I merely flattered myself) effect in Darwin a loss of confidence. It was evident that Max, Darwin, and apparently some others I had yet to meet were all aligned against me on the question of going ahead with humans. I did not believe I could stop them now even if I wanted to, and I realized I did not want to stop them because by this time too much of myself was invested in the ultimate goal. They had me, though at the time I think I fell back on an old idea—that I was merely an observer. It was my place to report, not to guide. Indeed, by virtue of being placed outside their inner circle and given no say in the decision to leapfrog the animal experimentation, the idea that now I really was merely an observer gathered some credibility in my own mind. I could feel that I was absolved of at least some responsibility.

At any rate, at the end of our conversation that day, both Darwin and I felt considerably more at ease with each other than we had ever felt before. Darwin clearly appreciated my not making a major issue of his heresy in bypassing the animal work and was, thereafter, far more open and frank with me. Now he had no objections if I stuck around after the holiday. He told me, with some awkwardness (for things were never to be entirely relaxed between us), that he had in mind to spend some time, while the others were away, with a girl friend. I almost had the brass to ask him if she was one of his egg suppliers. Instead I simply accepted his suggestion that I visit some nearby islands that he characterized as particularly beautiful and quiet. He would make the arrangements.

And so on Christmas Day I found myself in what struck me as the appropriately unusual position of feasting from a picnic basket, on a particularly fine stretch of uninhabited white sand beach, on roast turkey and plum pudding flown in from England.

CHAPTER 19

The hospital was quite a different place after the holiday. It served also as an outpatient clinic and it was a-buzz with customers from morning till night. Women with special gynecological problems, those interested in contraception, or those beset by real or imagined fertility problems were all referred to a female doctor I will call Mary who reported, in turn, directly to Darwin. From these women came some of the egg donors and surrogate candidates.

Mary, Darwin told me at the outset, knew everything. So did Paul. Mary was an M.D., Paul a young Ph.D. The latter, Darwin said, had come highly recommended by Darwin's shadow colleague. Mary was a biological researcher and surgeon; Paul's expertise was in cell biology and biochemistry. The nurses and technicians were all in the dark, Darwin assured me, as to the real nature of his work. They knew he was experimenting with eggs but only, they thought, to research various aspects of fertility and infertility and to explore possible new approaches to contraception based on egg transport and implantation.

Mary was a bit standoffish. Darwin told me that she regarded me as the "enemy." Paul was not exactly outgoing, either. He had a demeanor that was serious to the point of being almost grim. He told me that he was doing this work because it was paying well, it was good experience, and he was thus able to repay some of the favors his mentor (Darwin's absent colleague) had bestowed upon him. Neither Mary nor Paul saw anything immoral about their work. It was scarcely worth discussing in that regard, so far as they were concerned. Darwin, I could see, had chosen well from his point of view—for efficiency and not philosophy. With the likes of Paul and Mary about, he once told me, he could still hear himself think.

I was also told more about Roberto. He was in his thirties and was given to wearing flashy clothes and ostentatious rings. He had worked for Max locally for some time, I was told, and seemed to be something of a procurer. He would go through the factories and farms and invite various girls to come to the clinic for examination as possible candidates in a "study." So many, it seemed, succumbed to his blandishments that at one point, Darwin said, he was flooding the place with beautiful girls. Things reached such a pitch that Mary had become quite incensed, temporarily calling a halt to Roberto's solicitations while she lectured him on being more discriminating in his choices. She then ordered him not to bring more than four or five girls to the clinic in any given week.

Darwin hinted that Mary had "a thing" for Max and claimed that she had once told him that any girl Roberto recruited would by her very nature be unsuitable for the task of bearing Max's clone. Darwin obviously thought that Mary would like to be the surrogate herself.

The idea behind Roberto's work was to build up a dossier of surrogate candidates. The girls would come to the clinic, undergo complete physicals, and have their medical histories recorded and their photographs taken. Max, for whom women of this particular region held a special attraction, would examine these dossiers himself, discarding some, retaining others. He was being the more fussy, Darwin said, because the woman or girl who was selected as surrogate might become Max's mistress, or at least his principal mistress, and in that way would also be the "mother" or at least the maternal influence in the rearing of the clone.

Those candidates who passed muster physically were then

called back for tape-recorded interviews conducted by Mary. Max audited these tapes (he had a good enough command of the local language not to require an interpreter, though Mary, less fluent, often did) and made further eliminations. The women were paid for their visits and told they would be notified when the study began, provided they were selected to participate further. They were told only that this study had to do with health, both physical and mental, in the local area.

This might not be the most high-minded of operations, but at least it lacked truly sinister dimensions. There were government-approved and funded "research" programs I'd poked into that made Darwin's operation look saintly. On balance, I had to concede that the women who came to Darwin's facility seemed to benefit, not only in terms of the payments they received—and the importance of those could not be discounted, given the poverty in which many of the women lived—but also in the sense that they also received the best health care available in the area.

What I really feared was the possibility that Darwin and Mary might be performing surgery on some of these women *only* to obtain eggs, implant eggs experimentally, or carry out other procedures of benefit only to themselves. I spoke very frankly to Darwin about these fears. As usual, he overreacted. They were not "butchers," he said. For that matter, there were more valid indications for abdominal surgery among the women who came to the hospital clinic than he and Mary could possibly handle. In addition, many women of eighteen or nineteen already had two, three, or more children and were eager for tubal sterilization. Vasectomies were feared by the male population, and the Pill was economically and logistically unsuitable for the female population of the region. The tube-tying or -blocking operation was the most practical. It was only in the course of carrying out such procedures, Darwin said, that eggs were obtained.

He said he would let me watch one of his egg-gathering operations, but in the meantime he had a "surprise" for me. He took me to the maternity ward of the hospital, by far its busiest

125

department. In a small room near the nurses' station was an abdominal decompression unit or "suit" of the same sort I had seen in South Africa.

A young woman was in the unit, surrounded by several giggling, pregnant patients and a nurse, who was apparently instructing the women in the use of the apparatus, principally how to read and control the pressure gauge. The laughter intensified with our arrival. Darwin said a few words to the nurse, who made some explanation in the local language. Soon the women focused again on the patient in the suit. As the nurse proceeded with her demonstration, Darwin told me that he had become a convert to decompression after Max insisted upon obtaining a unit for the hospital. It was working so well, in fact, he said, that three more would soon be set up. They were simple to construct; also, he said, he had made a few "improvements" on the basic design himself. Darwin was as proud, I knew, of his engineering abilities as he was of his surgical skills.

Since at first he had regarded Max's prodding interest in decompression as something of a nuisance, he had even let the unit remain in its crate for some time after its arrival—until Max himself unpacked it on one of his trips and persuaded Mary to use it. When Mary admitted that she was getting "interesting" results, Darwin said, he began to pay heed.

I had collaborated on a book with Dr. O. S. Heyns, the "father of decompression," in South Africa.[12] I had also visited the decompression clinics there and had talked with many of the women who used the technique. I was delighted that it was being used at Darwin's hospital and told him I was confident that it would help him maintain pregnancy when he succeeded with his nuclear transfer. (I was always careful in such conversations to say "when" and not "if.") He said he was inclined to agree and wanted to know if I really thought it could enhance the intelligence of the newborn. I said that, while this hadn't been proved in the strictest scientific sense, the anecotal evidence as well as some objective data suggested that it could optimize IQ by increasing oxygen supply to the fetus during pregnancy. Anyway, I felt that its other effects more than mandated its use. I was sorry it still wasn't being used by more than a handful of doctors outside of South Africa.

Darwin summoned Mary and a couple of nurses who spoke English and asked me to tell them about my conversations with Dr. Heyns and my experience in observing decompression in use in South Africa. I started out a bit nervously, with Mary glowering at me, or so I imagined. I warmed to my subject as I spoke of Dr. Heyns (now deceased), who had been a scientist of the old school, the kind who knew quite a bit about the world outside as well as within the confines of his specialty, obstetrics and gynecology. I recounted pleasant days spent with him during his retirement, chasing butterflies, of which he had a Nabokovian knowledge, and, on our walks along the beaches of the Indian Ocean, identifying shells and various sea creatures. Heyns had studied medicine in England, was a Fellow of the Royal College of Obstetrics and Gynaecology, and had served as head of the Department of OBGYN at the University of the Witwatersrand in Johannesburg. He was a kind and humane man whose patients were as often black as white.

Heyns and some collaborators were doing research on the nature of uterine contractions, trying to fill in the gaps in our knowledge of the mechanics of labor. This was a difficult task because the powerful abdominal muscles influenced the contour of the uterine wall to the point where it was almost impossible to get a separate, isolated picture of the uterine activity itself. Various exotic sensing devices were used to pick up electrical impulses from the uterus, but there was always the possibility that some of these impulses were being generated by the abdominal muscles. In the course of these studies, Heyns and co-workers serendipitously discovered the usefulness of decompression in pregnancy.[13]

As he and his colleagues began thinking in terms of isolating the abdomen from external atmospheric pressure, the idea of a decompression chamber sprang into being. By reducing pressure

127

over the abdomen to well below normal atmosphere, Dr. Heyns reasoned, the intestinal gases within would surely expand, pushing the abdominal wall upward. He tried the first experimental unit on himself and showed that his hypothesis seemed to be correct; the abdominal muscles did extend upward and remained lifted for the duration of decompression. No ill effects were noted, even after prolonged decompression at atmospheres reduced to far below what would soon prove adequate for dramatic results in pregnancy.

The apparatus that was before us, I said, gradually evolved with further experimentation. When the first unit was applied to the abdomen of a pregnant woman having her first baby, she delivered in a remarkable three hours. (The average for a first baby is fourteen hours.) The second woman delivered in two hours and twenty minutes and with so little pain that no anesthesia whatever was required. Decompression lifted the abdominal muscles up off the uterus and let it work to its full capacity. In the literally thousands of cases that had since followed, I observed, the same excellent results had obtained. With decompression, anesthesia was rarely needed.

As the South African experience grew, other, even more important benefits of decompression emerged. By using the units well in advance of delivery, even several months, the doctors found that fetal circulation and oxygenation could be optimized and complications of pregnancy resulting from toxemias, placental insufficiency, and the like all but obliterated. Thus were the chances of miscarriage and fetal abnormality also minimized. These were the effects Darwin was particularly interested in where our project was concerned.

I mentioned the numerous scientific papers that had been published on decompression in such prestigious medical journals as *Lancet,* many of them documenting the lifesaving benefits of the technique, others documenting its various complex effects on fetal and maternal circulation.[14]

Mary seemed disarmed by my lecture and even smiled at me

as we broke up. I asked her if I might chat with her at some point before I left, and she said yes. I was eager to find out more about her surrogate program and how she assessed Darwin's progress to date and his chances of success.

CHAPTER 20

The next morning I was an observer in surgery. Darwin, with Mary, Paul, and an anesthesiologist standing by, prepared to recover some eggs from a young woman. As a benefit of this operation, I was told, she would be sterilized via tubal electrocoagulation. Darwin explained that, though it was possible to culture and ripen immature eggs obtained, for example, from the ovaries of women who were undergoing complete hysterectomy, much better results were achieved when eggs were permitted to grow almost to maturity in vivo—in the body.

To speed up the process and obtain more eggs, patients like the one about to be operated on were given the fertility drug Pergonal, consisting of pituitary hormones that promote development of multiple follicles and ova (instead of the normal single ovum each month). Darwin assured me that this artificial induction of extra eggs was not about to deplete the ovaries. The typical woman is born, he said, with about half a million egg cells. Yet only about 500 of those would mature in her lifetime. It made you wonder, Darwin said, what the other 499,500 were for. Maybe nature foresaw something we didn't. Maybe, he joked, or at least I think he joked, nature was providing for "egg-snatchers" like himself.

The Pergonal had to be injected at the proper time, generally on several alternating days early in the menstrual cycle. These injections were followed by the administration of what Darwin called "the kicker"—a single injection of another hormone (human chorionic gonadotropin) which put in motion the final chromosomal maturation process within the egg and caused the actual ovulation or extrusion of the egg from the ovary. In the natural course of events hormones similar to the ones he used were released (in lower quantities) by the body, and ovulation occurred generally about thirty-six hours after the kicker hormone had "kicked in." The object, then, was to recover the eggs sometime *after* the maturation hormone came into play and *before* the fully matured eggs left the ovary and started down the fallopian tube on their three-day journey to the womb.

To get the eggs, Darwin scheduled surgery a few hours prior to the time at which ovulation should occur—more than thirty hours after he had injected the human chorionic gonadotropin. He made a small incision just beneath the navel and inserted through this aperture the long, slender laparoscopic viewing device which illuminates the peritoneal cavity with a fiber-optic light source. With this he was able to locate quickly the blisterlike follicles on the surface of the ovary that contained the almost-ripened eggs. The eggs were "plucked" from these blisters with a needlelike suction device that was pushed through the lower abdominal wall and guided to the targets under direct laparoscopic visualization. This time, Darwin got five eggs, which he immediately turned over to Paul, who placed them in a culture dish containing a specially prepared growth medium. Darwin then proceeded to block the tubes of his patient with a miniaturized electrosurgical grasping forceps. The whole procedure took little more than half an hour. The puncture wounds in the patient's abdomen, I was told, were self-closing, and the young woman would be out of the hospital the following day. Although she had been given a general anesthetic, these new laparoscopic procedures, Darwin noted, *could* be performed with only local anesthesia.

Darwin had been obtaining eggs in this way almost from the beginning of the project. Their fates were complexly varied.

Many were studied for their reactions to and behavior in a variety of media, which included a number of chemical, physiological, and even viral ingredients. They were subjected to all manner of environmental variables of alkalinity, acidity, osmotic pressures, and ionic and atmospheric factors. Surface tensions and internal pressures were gauged in the course of microinjecting and otherwise micromanipulating them. Extensive efforts were made to determine the moment in their maturation at which they might be optimally vulnerable to "fertilization" by body-cell nuclei, accomplished either by fusion or by microsurgery. They were chemically primed, stretched, denuded of their outer protective layers, pumped up, cooled down, injected, dissected, assembled and disassembled, fused, fertilized, jolted, shocked, irradiated, and sometimes completely dissolved in experiments aimed at getting them to give up their own nuclei and accept the innards of a "foreign" body cell.

Some were fertilized by sperm in vivo and then washed from the fallopian tubes; others were fertilized by sperm in test tubes and then enucleated—deprived of their nuclei—by chemical or surgical means while still in the single-cell stage. Still others fertilized by sperm were permitted to become multicelled embryos and then dissected into individual cells and enucleated. What was left of the enucleated egg was the cytoplasm (figuratively, the white without the yolk). The aim was to get by hit or miss a fully switched-on cytoplasm which, with its monitoring/instructing mechanism, would throw a body-cell nucleus into gear and keep it there. No possibility was being overlooked, Darwin said.

All this said nothing of the body cells. Had he reached any conclusions about those? Which type or types might lend themselves to this procedure? He had tested several kinds, he said, under different circumstances. Some cells were slow to divide, and obviously, he wasn't going to concentrate much effort on those. Brain cells, for example, apart from their relative inaccessibility, were in no wise tempting, given their undividing, static status. What were needed were cells with rapid turnover. Embryonic cells (in which differentiation was not yet complete) would be ideal but, alas, Darwin lamented, his client had long since left the "larval state" behind.

I asked another question or two on body cells, but it was finally evident that he was not going to tell me at this time which cell he regarded as his best bet. This, he said, should not be revealed. If the identity of the clone were ever "betrayed," he said, it would not do to have this unique individual known as one who had "started life as a piece of bone marrow, a blood cell, or a scrap of cancer." That, he said, would add to a stigma that might already be overwhelming, depending upon "the mood of the times."

I didn't buy this at all. Instead, I concluded that it was natural for him—indeed, for any researcher in his position—to want to protect his research, or at least many of the specifics of it, until such time as he felt he could claim it as his own by publishing its details in an appropriate journal. I was convinced that Darwin, if he succeeded, would ultimately reveal his own part in all this. His ego would not permit him to do otherwise—though he might have to wait some time, for he would not take lightly the consequences of Max's or his colleagues' wrath.

This same ego, I thought, played a large part in his ambivalence as to how much to tell me. There were times such as just now when he was very self-protective, but in many instances it seemed that before long he would be making a point of letting me know the details of whatever it was he had previously refused to divulge. For example, one day it had suddenly seemed of no concern to him that I should learn he had performed a biopsy on Max for the purpose of obtaining liver cells. Again, he was quite open about working with erythrocytes (red blood cells) and epithelial cells taken from the lining of Max's mouth. (He was to say later that the secrets were in "very little things anyway—like patience and the way I lubricated my eggs." I couldn't tell whether he was joking about the latter or not.)

It was as if, since he knew I would be writing about this work

132

no matter what, he wanted to make sure I didn't overlook any of his accomplishments. What it boiled down to, I thought, was that sometimes his ego overwhelmed his better judgment. I was faced then with having to decide whether I could ethically divulge everything he told me. I decided that in a few instances I could not—or should not.

CHAPTER 21

During my two weeks at the lab I watched a number of experiments with eggs and embryos. If Darwin and his collaborators should ever succeed in converting a body-cell nucleus into an embryo, they still would be faced with the problem of getting that embryo to grow to the point at which they could implant it into a surrogate womb. Human embryos have to reach the "blastocyst" stage, usually at about five days' growth, before they develop the trophoblast cells that enable them to attach to the lining of the womb. These trophoblasts are like little claws that literally dig into the uterine wall and enable the parasitic embryo to suck in enough sustenance to keep itself alive and growing.

In the normal course of events the female body provides the ideal biochemical environment for the growth of the embryo, both before and after it attaches itself to the lining of the womb. But a test-tube embryo, an egg fertilized by sperm or implanted with a body-cell nucleus in a laboratory container, must be artificially nurtured—at least up to the early blastocyst stage. Thus, before Darwin could even think about a successful cloning, he knew that he would have to master the test-tube requirements of the embryo. He would have to figure out how to make it grow.

There were three different ways by which he could obtain embryos for his experiments. He could artificially inseminate some of his patients, perform surgery, and try to recover a fertilized egg from their tubes. That would be time-consuming, cumbersome, and risky. He could also try to create embryos through the cloning process—by implanting body-cell nuclei into eggs. I have described in the preceding chapter the variety of efforts he was making to prepare the cytoplasm to switch on the body cell, but since he was still not having a great deal of luck, this was not the most practical approach, either. He concentrated instead on the third option—the creation of embryos through the test-tube fertilization of eggs with sperm. In this effort he had been quite successful, and this alone was an accomplishment worthy of note, for test-tube fertilization of human eggs had been achieved only by a small number of researchers around the world—men like Steptoe, Edwards, Shettles, and Bevis.

The objective of all this, once again, was to learn how to keep an embryo alive and growing in the test tube once you'd "constructed" one. If you could master this second step in the overall cloning process, then you'd be ready to move right on to the third step—implantation of the embryo into the uterus—just as soon as you achieved the first and possibly most difficult step: producing a clone by "fertilizing" an egg cell not with sperm but with a body-cell nucleus. It made sense, from the point of view of efficiency, to concentrate heavily on steps two and three even before you had mastered step one.

I watched as three eggs that had just been taken from the ovaries of a young woman were dropped into a culture dish containing a combination of natural and synthetic ingredients. These included such things as sodium, calcium, bicarbonate, other inorganic ions, fatty acids, proteins, sugars, and so forth, and also blood serum and a bit of fluid from the ovarian follicles.

The "fine tuning" of growth mediums like this one, Darwin indicated, was of prime importance in determining whether the eggs would mature, be fertilized, and ultimately attach themselves to the lining of the womb. He and Paul had experimented with dozens of different media and still felt there was room for improvement.

134

Once in the culture dish, the eggs were protected from contamination by a layer of silicone oil (inert so as not to affect the culture medium or the eggs) which floated on the surface. This culture dish, however, was only part of what amounted to an artificial womb, for Paul next placed it in an incubating chamber that looked something like a built-in oven.

Here it was possible to do even more fine tuning, precisely controlling the temperature, pressures, and atmospheric conditions. You tried to reproduce, as closely as possible, the conditions that normally prevailed in the tubes and womb right after conception. It was almost like studying and then controlling the weather, Darwin said. Even such things as the high relative humidity of the oviducts had to be taken into account and duplicated. So did oxygen tension and the composition of carbon dioxide, oxygen, and nitrogen. Temperature had to be carefully maintained at the proper points. The pH (a term used to express acidity and alkilinity) was kept invariably alkaline, but only slightly so.

The solution in which these three eggs were now incubating was called the "maturation medium." It was in this medium that the eggs would complete the internal development which would make them ready for fertilization by sperm. These eggs were obtained thirty-one hours after Mary had injected their donor with the hormone shot that promoted ovulation. Almost invariably, it was thirty-six hours after this shot that the eggs were fully ripened and ready to be exposed to sperm. Thus they were incubated in the maturation medium for five hours.

Before these five hours were up, Darwin prepared the sperm. I felt a little foolish but asked him where he had got it. He said that when he had first begun these experiments he had used his own. Later he had, as he put it, "imposed upon Paul" on a few occasions. Finally, the useful Roberto was called upon to recruit some sperm donors, an effort that caused a minor sensation among the locals, who were apparently amazed that they could earn a day's salary in a few minutes and with so little effort. It wasn't long until the "locker," as Darwin called his frozen-sperm bank, was well stocked.

Darwin explained that sperm have to undergo a process known as "capacitation" before they are capable of fertilizing eggs. Parts of their "heads" have to be unsheathed by chemical constituents of the female reproductive tract. Capacitating agents were included in the fertilization medium in which eggs and sperm were finally joined.

Paul "washed" the sperm first—put them in a solution that rid them of excess seminal fluid—then placed them in the fertilizing medium, which was more highly alkaline than the maturation medium. The eggs, sucked up out of their bath with glass pipettes, were placed in the fertilizing medium after the sperm.

Darwin said that all three eggs would probably be fertilized within four hours. They were left undisturbed, however, for nearly twelve hours to ensure that fertilization was completed. What he then assumed were one-celled embryos were transferred to another medium—this one designed to promote cell division and growth. The composition of this medium, Darwin said, was of great importance in determining the "staying power" of the embryo in the critical stage immediately after it was implanted in the uterus. Consequently, he had devoted a great deal of time to studying the requirements of embryos in the earliest stages of pregnancy; and to impart that staying power best, he had concentrated in particular upon experiments with different energy sources for his media, such as various sugars, pyruvate, lactic acid, and so on, with increasingly better results.

It was Darwin's view that most of those who were attempting human embryo implants and transplants had failed because they had tried to implant too early. Steptoe and Edwards, for example, were trying to get a pregnancy with 16- and even 8-cell embryos—long before nature was ready to accept them, Darwin felt. One reason for the rush was that some researchers were unable to maintain their embryos in vitro much beyond the 8- or 16-cell stage or feared that if they did so the embryos might be damaged.

But it simply wouldn't do, Darwin insisted, to hurry things.

It was necessary to come up with culture media from which "fully energized" embryos at the "morula" (32-cell) or, better yet, "blastula" (64-cell) stage could emerge. Darwin repeated the words "fully energized" again, for it was possible, he said, to obtain from culture blastocysts which appeared grossly normal but which were actually "anemic" and incapable of holding on to life in the womb for more than a very brief period.

To prevent this as the embryos grew in culture, he had made various useful modifications in the media, with some "vital inputs" just before implantation. Until very recently he too, he admitted, had been working with 8- and 16-cell embryos and was only now arriving at a point at which he expected to get dramatically better results. He was already able to grow embryos to the blastula stage and hoped soon to have them beefed up to the point where they would have the endurance to survive the implantation effort.

Darwin then revealed that he had already tried implanting some of his test-tube embryos. This alarmed me a bit. I asked him what he would do if he got one of his subjects pregnant through this procedure. Were all the women in whom he was implanting embryos infertile (due to blocked tubes, for example) and desirous of having children? Again, were his subjects being fully informed of the risks—one of which might be unwanted pregnancy? He was a bit evasive, which made me think the worst.

At any rate, he said that he was working only with married women, so at least no virgins would suddenly find themselves pregnant after participating in one of his "studies." And yes, he was working with infertile women where possible, he said. Were these the same women *he* had rendered infertile with his tube-blocking operations? And how did he know that he was not potentially impregnating some of these women with eggs which had been fertilized by the sperm of their own cousins, brothers —or worst enemies? I wronged and underestimated him with such suggestions, he said. Precautions were of course being taken against implanting embryos of incestuous origin.

So far, in any event, there had been no lasting "takes," by which he meant pregnancies, though of late he claimed to be getting encouraging signs. Analyses of hormone levels in the urine of one woman had indicated for a short time that she had become pregnant after an embryo implant.

I was permitted to watch one of these implant operations. It was a relatively simple procedure, at least as Darwin and Mary did it. There were two methods that could be used. One involved making an incision in the abdomen and the uterus and implanting the embryo right through the wall of the womb. This approach was not favored by Darwin; he found it "almost brutal" and, in any event, unnecessary. He preferred the method advocated by Shettles and later adopted by Steptoe and Edwards.

The blastocyst was sucked up from the growth medium into a syringe which was next attached to a long rubber catheter. Mary inserted this into the woman's vagina and right on through the cervical opening into the womb. Then Darwin allowed the embryo and a small amount of protective fluid to fall, under the force of gravity, out of the syringe, through the catheter and into the woman's uterus. The woman was told to remain motionless for several minutes thereafter.

Real progress, it seemed, was being made with steps two and three—embryo culturing and implantation. But now back to step one. After all Darwin's work with egg cells, what major problems remained, I asked, before he could activate an egg by the implantation of a body-cell nucleus in the same way that sperm penetration would activate it?

"Activation" is a technical term that refers to the final maturation of the egg and the process whereby its cytoplasm instructs the nucleus to start dividing and create a whole new individual. Until an egg is fertilized by a sperm cell, it is in a state you might call animated suspension, arrested in what is known as the "metaphase II" stage of cell division. Fertilization results in a number of events that lead to a resumption of cell division. As to body-cell activation, Darwin said, this event could be precipitated either by

138

the microsurgical injection of a body-cell nucleus or by the chemical fusion of an egg's cytoplasm and a body-cell nucleus. Simply lowering the temperature in which the egg was incubating could sometimes effect activation.

A far more significant problem, he said, was to get the egg-cell cytoplasm and the body-cell nucleus synchronized—that is, dividing at the same rate. Egg cells divide rapidly, whereas most body cells divide rather slowly. Chromosomal breakage could develop if the body-cell nucleus were forced by the cytoplasm of the egg to try to divide before it was ready. There were ways of trying to get around this problem. One way was to experiment with various of the more rapidly dividing body cells. He had found that, when he grew body cells of different types in cultures of varying constituents and atmospheres, he could sometimes speed up or slow down the rates at which they divided. Even without any of this "tuning," cells grown in culture divided more rapidly than they did in vivo, or even immediately after being removed from the body, and this was advantageous.

That was only one of the possibilities he was experimenting with, he said. There were ways of stopping cell division altogether, then starting it again at the desired moment. There was also some indication that cells taken from blastocysts in which cleavage abnormalities were evident might provide "the ticket," as he put it, for reasons detailed in chapter 27.

This talk prompted me to ask him something which seemed a bit outlandish but which Darwin, nonetheless, soon indicated he had not overlooked. Max, I knew, had a skin cancer that had been yielding to treatment and was of the sort that is 100 percent curable. The similarities between embryonic cells and cancer cells had often been noted. Was it conceivable that one of those rapidly cycling cancer cells might be suitable for the task at hand? Or would there be a danger of perpetuating the malignancy? Probably not, Darwin said; amphibians had been successfully cloned from what had been identified as adenocarcinoma cells (malignancies arising in glandular epithelium) and from kidney tumors.[15] Other experiments were in progress in which cancer cells were being injected into early embryos of mice. Thus introduced, they proliferated throughout the bodies of the mice, expressing themselves in almost all of their tissues—not as malignant but as normal cells. When "recycled" in this way, it seemed, they reverted to complete normalcy, indicating that the original cancer from which these normal cells arose was due not to an inherited, mutational change but rather to an aberration of gene expression which was reversible and might have been caused by any number of environmental factors.[16]

There was a Nobel Prize waiting for the researcher who could "put handles" on the control mechanisms of a cell, Darwin asserted. Proteins, hormones, and other complex substances were known to pass from the cytoplasm into the nucleus at various stages in its development, and many interesting results could be obtained by injecting small amounts of cytoplasm from one cell to another. Eventually, he thought, someone would find the "master molecule" that could be added to a cell—any cell—and cause it to "dedifferentiate" completely; that is, to lose its specialty, such as being a retina cell or a heart cell. Then it would behave like a fertilized egg cell, which could develop into a whole new individual.

When that time came, he added, it might not be necessary to go through the sticky task of nuclear transfer. A cell switched on by means of this molecule or combination of molecules might simply be nourished in the test tube with the proper energizers and then be implanted into a surrogate uterus at the appropriate time. Other molecules or combinations thereof would also, he was confident, be discovered by which cells could be induced to dedifferentiate only partially, in predictable and controllable ways. For instance, they might be induced to commence dividing to create not whole new individuals but simply whole new livers, hearts, pancreases, or even brains. Thus, when you needed a spare part to replace a damaged, diseased, or worn-out one, you could have the part grown from your own cells and have it surgically implanted without fear of rejection. It sounded wild, Darwin said, but he was certain it would be possible one day.

CHAPTER 22

Mary's office, particularly in contrast to the clutter of Darwin's cubicle, had a severe, almost antiseptic orderliness about it. It was not made more cheerful by its lack of natural light. The jungle pressed right up against the window. The only effort to make the room more homey was evident in a large poster on one wall—a picture of a spectacular snow-capped peak. Mary said she had grown up in sight of this mountain.

I guessed Mary to be about fifty. She was a large woman, but statuesque rather than portly, the sort of woman many would call handsome. Her thick black hair with its flecks of gray was particularly striking. Her general demeanor was of the no-nonsense category. Her voice, however, was surprisingly soft. Darwin told me that she had been widowed many years earlier and had never remarried.

I learned in my first conversation with Mary that she had a background in psychiatry as well as in obstetrics and gynecology. She had worked in this geographical area before and had some knowledge of the local people and language. She would say nothing about how she came to work with Darwin. I assumed that they had worked together before. (Later I got the impression that it was Max she had known for some time. I was never sure.)

Mary seemed intent upon educating me a bit or persuading me to see things in the proper light, which was to say from her perspective. She acted a bit bored with my questions about cloning. I must understand, she said, that cloning was mere busywork compared with many of the things that were being accomplished in the realm of molecular biology. The "controversial" work in which the genetic material of unlike species was now being spliced together, in some cases to produce new life forms with the capacity to do previously unheard-of things, interested her a great deal more. I soon realized that she was talking about recombinant DNA work, which some months ago I had characterized to Max as "alarming" in some respects. This research, she said, defined the real leading edge of genetic engineering. By comparison cloning was "a sort of oddball entity" in a congress of laboratory wonders.

Mary's position on recombinant DNA could be summed up in a few words: she was for it. "Mother Nature" was not the kindly element some romanticized "her" to be. It was only by constantly "interfering" with things that we had got to the top of the evolutionary ladder—and we weren't likely to remain there much longer, judging by the history of other species, if we didn't go right on interfering. If anything, she said, we must step up our interference. In three billion years more than one hundred million species had lived on the earth. All but two million of those were now extinct, she said darkly, and there was no reason to expect that those two million had any special blessing of nature. Their days, too, were numbered. It was only a matter of time until they were eliminated—supplanted by something better, or at least more adaptable. We humans were the only one of those two million species with the ability to beat nature to the punch and do our own bettering.

She said that the kind of cloning she and Darwin were out to achieve was easy, compared to the cloning of genes that was now not only contemplated but in some cases already being done in laboratories around the world. She marveled at the progress which had been made in the field of molecular biology since the 1950s and said that the middle decades of the twentieth century would ultimately be recognized as the most significant period in the evolution of mankind. Such things as the discovery of fire, nuclear energy, the development of the wheel and various tools —even these would pale alongside the recent developments in genetic engineering. We were still too close, she said, to appreciate fully the meaning of these developments.

It had previously taken five to twenty million years for a single mutation to become completely incorporated into a species. Now, Mary said, we were on the threshold of being able to effect even radical mutational changes in the genetic schemata of any given species practically overnight. The year 1973 in particular, she believed, would be remembered for its "historic breakthroughs," those that enabled researchers to splice together the genetic material of virtually any two organisms they wanted, creating what one scientist had called "DNA chimeras" (because, he said, they were "conceptually similar" to the mythological Chimera which had the head of a lion, the body of a goat, and the tail of a serpent.)

The breakthroughs that Mary spoke of had enabled scientists to transplant individual genes into bacteria which reproduce through simple division, creating carbon copies of themselves each time they divide. By this ingenious method it was possible to make billions of copies of the transplanted gene in a few hours. This technique, which was accomplished through several neat tricks of genetic surgery, was called "molecular cloning." In place of implanting a whole body cell into an enucleated egg cell, scientists were already successfully taking individual genes and splicing them into incredibly tiny strands of bacterial DNA, where, through the reproduction of the bacteria involved, they were replicated millions and billions of times over.[17]

The idea was not to confer bizarre new properties upon bacteria but to analyze the function of individual genes (of which, in the human, there were hundreds of thousands) and thus finally arrive at the most basic, molecular foundation of disease and abnormality. It was also possible, through these recombinant techniques, Mary said, to engineer bacteria that would, after you spliced them into the appropriate genetic instructions, turn out large quantities of badly needed enzymes, hormones, antibiotics, and other chemicals needed to study or treat disease. Already molecular biologists were at work to create a bacterium that would produce the insulin that diabetics so badly needed.[18]

The new techniques also enhanced our ability to isolate and even construct new genes for specific purposes. Eventually some of these genes might be introduced by one carrier or another, or possibly by viruses, into individuals deficient in them, then activated and caused to reproduce themselves.

Of course, it was also possible to create new life forms with these new technologies, and this was where so much of the controversy came in. There was the fear—not entirely unfounded—that scientists might inadvertently create a bacterium or virus with virulent new properties against which we might not be able to defend ourselves. Unscrupulous governments, even terrorists, might engineer such malignant mutants—after having also engineered "antidotes" to which only they would have access. The recombinant work had many powerful critics, some of them Nobel Prize winners, and they were able to paint horrifying scenarios of man-made life forms seeding disease, disaster, and doom. In time, they would also be able to point to some actual close brushes with disaster.[19]

On the other hand, those who favored the new work saw an enormous potential for good—in the creation, for example, of bacteria with the capability of "eating" oil spills (proposed by General Electric), converting waste products into usable energy sources, and the like.[20] There was even on the drawing boards a recombinant plan by which various plants could be induced to obtain nitrogen directly from the atmosphere, obviating the need for costly and environmentally harmful fertilizers.

Beyond this, Mary said, given the fact that we were now capable of assembling and disassembling cells, of isolating and synthesizing genes, and of overriding the natural instructions of a cell nucleus with instructions of our own devising, there was no reason why we could not now begin to contemplate the ways in which we might remake ourselves—in an image that only our own minds, our own hopes and dreams, could mirror. These, Mary acknowledged to my relief, would be the most important decisions we ever made. No doubt some would argue that no such decisions should *ever* be made; but the will to survive, she was confident, would carry the day, even if it did require us to "play God," as some would have it.

She claimed that it might be necessary to play God, but I said I would be terribly nervous to be in the company of anyone who volunteered too quickly for the part. To be truthful, I was a bit put off by the breathy, hortatory tone of all this talk and inclined to think Mary was overestimating man's "new abilities." With the passage of time, however, I would find even such sober journals as *Scientific American* broadcasting the same tone. A 1977 report in that magazine detailing some of the latest breakthroughs in recombinant DNA work was to conclude that, with chemical synthesis and cloning technology, it was now possible for the molecular biologist "to fabricate any DNA sequence he wants, to produce it in unlimited amounts and finally to insert it next to any other piece of DNA he chooses." And the bottom line managed even at that late date to chill me a little, perhaps because it was so explicit: "For the first time, man has developed the capacity for almost absolute control over the material of the genes."[21]

Mary observed that Max was very enthusiastic about the recombinant work. I had heard him talk about this, too, and I asked her if she didn't fear private entrepreneurs' going overboard and possibly endangering us all in their pursuit of profits. (Among other things, Max was interested in cloning whole forests and in redesigning the human digestive system so that mankind could henceforth comfortably eat grass and hay—a scheme a major corporation, I subsequently learned, took seriously![22]) I soon learned that Mary was a firm believer in unbridled free enterprise. She and Max, I thought, were a perfect match.

My own views on the new work were mixed. In principle I

145

supported it. The potential benefits were simply too enormous to foreclose all experimentation. I only hoped that the precautions being formulated would prove sufficient. In some regards, the guidelines then being formulated by the government's National Institutes of Health seemed barely adequate.[23] They did very little to ensure the prudent pursuit of recombinant goals among the purely profit-motivated. Who would monitor the private sector—which did not depend upon Uncle Sam for funding? Who would monitor the Maxes of the world?

I was surprised to find later on that Darwin was often as ready to play devil's advocate as I. Actually, he favored most of the new work, but it salved his conscience, I think, to call attention to some of the perils of recombinant research. What would Max think, Darwin asked him slyly, if some of General Electric's proposed oil-eating bugs got into the wing tanks of a jet in which Max was a passenger? Were we hereafter going to have to pasteurize all the oil we used in our planes, cars, and other conveyances? And how would Max feel if terrorists or political groups he didn't agree with concocted "recombinant bombs" using materials which could currently be purchased through the mail by anyone with a stamp? Even a bright high-school student, Darwin once said, might be able to put together his very own Andromeda strain. The details of plasmid engineering had all been published and were easy enough to come by.

For Darwin it was nice that the public, the press, the "religious fanatics," as he called them, and, yes, many of his peers now had in recombinant DNA research "a new whipping boy." Perhaps now things like in vitro fertilization, embryo transfer, and possibly even whole-body cloning would be seen for the "uncomplicated" entities they were. With a little help from a molecular biologist, Darwin quipped, a normally benign bacterium called *E. coli* could clone a menace that might lethally "infect the world in a matter of days or weeks." The most *he* could manage, Darwin said, would be to clone a man who might, over a lifetime, "turn out to be a pain in the ass."

146

CHAPTER 23

At the end of my first conversation with Mary she asked me where I was staying. I told her that Darwin had arranged for me to sleep in one of the hospital's private rooms. She said that was wasteful and suggested that I stay at "the bungalow." The bungalow was Max's local residence, and was, in fact, quite a bit more than a bungalow. It was an impressive home with its own private beach and plenty of heavily wooded and carefully guarded grounds. It also had its own live-in servants. Mary had been staying there ever since she came to the area.

The place was big enough to accommodate Darwin and Paul as well, and I thought it odd that they preferred their spartan quarters at the hospital. Darwin muttered something about a lady's needing her privacy, but as Max later explained, it was Darwin who was worried about privacy—his. He apparently feared Mary wouldn't approve of his occasional late hours or some of the guests he brought home. As for Paul, he preferred to be near the lab at all times, particularly since he often worked late into the night and was the sort who would sometimes start an experiment at three in the morning.

And so I moved out of the hospital and into the bungalow. We made the daily commute by white Toyota. Max had provided both Mary and Darwin with identical vehicles—"company cars," Darwin called them. Mary was not much more accessible at the house than she was at the hospital. She spent most of her free time working on a manuscript related to some research she had been engaged in for several years.

Darwin and Paul usually joined us for dinner, since the fare at the house was so much better than at the hospital. The vegetarian Max, in honor of his guests, had sent in a freezerful of Argentine beef. It was during one of these dinners that I finally got Mary to talk a bit about the surrogate-recruitment effort. It began with her response to some comment I had made about the prospects for success.

Again she said that cloning a man was relatively simple, and the conditions for success were favorable at this facility. The team had all the funding they needed, privacy, no red tape, and plenty of "raw material," by which I guessed she meant experimental subjects. When I questioned the ethics of implanting a test-tube embryo into an unsuspecting patient, I found her even touchier than Darwin, who remained silent through most of this conversation. She denied that there had been any unsuspecting subjects. All those in whom implants had been attempted had been fully informed of the nature of the procedure. What about attempts to implant egg cells "fertilized" with body-cell nuclei? Would those women be fully informed? Would they be told that the offspring that might result would be a "clone"? Or what a clone was?

First, Mary said, no cloning effort had yet been made. All the eggs used in the implant attempts had been fertilized by sperm. Many of the women receiving these implants wanted to become pregnant. Others had agreed to receive the implants for a fee and were told that, if the implant held and they didn't want to remain pregnant, an early abortion could be performed. Surely, I told Mary, neither Darwin nor she would be very happy if the abortion option were exercised, for this would deny them the opportunity to see if fetal development would proceed normally. Nonetheless, Mary said tersely, the option was "available." Besides that, if anything seemed to be going amiss, an abortion would be performed without hesitation.

I asked about the "ultimate" surrogate—the woman who would be selected to carry the clonal cell—Max's. How would the selection be made? Mary said that Max's body-cell nuclei would be implanted only in the woman or women who had, after careful screening, been found suitable for the role. I wondered about the value judgment in this but held my tongue and let her continue. A great deal of effort was being expended to find such a suitable surrogate. She reminded me of her psychiatric qualifications and said that she had worked for a time as a therapist, primarily treating and counseling women with problems related to pregnancy, fertility, and motherhood.

In any event when the "right one," as she put it on several occasions, was finally found, an exception would be made insofar as the "ethical rules" were concerned. Mary seemed slightly uneasy as she began getting into the specifics of what I wanted to hear. The successful candidate, she said, would be single and a virgin. (I almost smiled.) "A bit of a ruse" would be required, she added, still uneasy. The surrogate would be told that she was being retained to assist a couple to have a baby; that the wife had lost womb and tubes to surgery, but was still able to produce viable eggs. and one of these would be fertilized by her husband's sperm in vitro. The conceptus would then be transferred to the surrogate womb. The couple, the surrogate would be told, insisted upon a virgin for religious, aesthetic, and medical reasons.

The more I thought about it, the more practical the ruse sounded. It did seem likely that a couple in such circumstances would prefer a single woman. For one thing, without a husband, the surrogate would not be as likely to want to keep the baby once it was delivered. The possibility of such a problem's arising could not be taken lightly; nine months was a long time and pregnancy an intimate process. Many a woman might say, after all that time and effort, genes be damned, she had borne the child and she would keep it.

There were other considerations that commended the choice of a single surrogate. She could be more easily isolated. A woman with a husband and children couldn't be expected to abandon them completely for nine months, and yet, to help ensure success, Darwin and Mary would want to keep the surrogate under constant surveillance. They would treat her, Mary said, like a "habitual aborter," meaning a woman given to frequent miscarriages. They would take every precaution to minimize accidental occurrences, bad nutrition, or health habits that might jeopardize the pregnancy. A married woman with responsibilities would be at much greater risk of accident, disease, and stress, physical and mental.

The virginity requirement I had a little more trouble understanding, at least in terms of its practicality. Well, Mary said, a virgin would be unlikely to try to claim the baby. And a virgin would be more manageable. How so? Once again Mary seemed uncomfortable, but pointed out that a nonvirgin might want to keep a boyfriend around or might in any case be difficult to keep confined for nine months.

All right, I said, but weren't virgins going to be difficult to come by? Well, not so difficult as in the United States, Mary said, smiling a little. Premarital sex was still frowned upon in the local culture. There was a problem in that girls married young, however, and it might be difficult to find an "acceptable" virgin who was older than seventeen or eighteen. Finding a girl of sixteen or seventeen with the requisite maturity to handle this whole procedure was not easy, particularly since Max was being very "picky." I pressed for clarification, and Mary said with some evident disapprobation that he was insisting upon a very pretty one. Darwin and Paul suddenly seemed terribly interested in their plates. Actually, Mary noted, the very pretty ones were the first to be married off, and this tended to narrow the market in "acceptable" virgins even further.

I thought there might be in Max's desire for a virgin, apart from the practical considerations, a dash of the old-fashioned double standard. Max had played the field far and wide for many years, and in that time, I had a hunch, had seldom demanded credentials as to virginity. But the surrogate would be the closest he had come to a wife. I guessed it might make a difference to him.

If Max was planning to take the surrogate as a mistress and install her, in effect, as the "mother" of his "child," why couldn't they simply inform the girl of what was happening? Mary said that Max wouldn't decide on the girl's future until he saw how she responded to the baby—and how she responded to Max, and he to her, over a period of time. A cover story was needed in the meantime. For that matter, Max might *never* tell the girl what had really happened; he might decide it was best, even if he did take the girl as a mistress, to claim only that his "wife" (the imaginary egg donor) had died. The surrogate, if she were fond of the baby and if she liked Max, would be the wife's obvious successor. It would, Mary felt, seem natural to her.

All in all, I decided it wasn't a bad plan. What were the chances of finding the "right" one? Mary shrugged, looking discouraged. They were working on it. She and Roberto? She winced at the mere mention of the name. They had been through dozens of possibilities, and only two girls were currently in the "active" file. This meant, Mary explained, that Max had not *yet* found cause to disqualify them. Both were in their teens and both were "very attractive."

I asked Mary what sort of questions she put to the girls in her preliminary screening. She often played it by ear, she said, but also asked questions Max himself had suggested. Questions about their hopes for the future, their thoughts on marriage and children, what sort of home and husband they hoped to have, whether they would like to travel, what kind of friends they had, what their families were like, whether they had boyfriends, whether they liked younger or older men, or if they thought much about men at all. They were also asked many questions from standard personality inventories that could be scored—questions relating to self-worth, self-image, and the like. Their personal habits were probed, such things as diet, smoking, drinking, and sexual practices if any. They were asked about their religious and moral beliefs, their cultural views; they were asked to describe "the most exciting experience" of their lives, "the most frightening experience," and so forth. They were asked to draw pictures of themselves, to sing (if they could or would), and to describe the activities they most enjoyed and most disliked. They were asked to "make up a lie," a request many of them had trouble not only in fulfilling but even in understanding.

Darwin cut in at this point to say that he certainly hoped

some surrogates were selected soon. Getting Max together with the right girls, he said, was beginning to look as if it might be more difficult than getting an egg together with the right body cell. There were beautiful girls everywhere in the area, he went on with enthusiasm; what was the problem? Turn the detail over to him and he'd have six girls the first day.

Mary made no effort to conceal a grimace. She could understand Max's wanting to be careful, she said. This was no casual matter. Roberto was the real problem; if she could get it through his head that Max was interested in brains as well as—she hesitated a moment and then said it—"boobs," maybe they'd get somewhere.

Darwin rolled his eyes upward, shook his head, and went back to his dinner. The phlegmatic Paul merely cleared his throat, and Mary, apparently piqued at having shown some emotion, also lapsed into silence.

CHAPTER 24

As my stay at the facility lengthened, it became apparent to me that Paul was at least as important to the project as Mary. Indeed, he seemed almost indispensable. Utterly humorless and seemingly addicted to his work, he was what some might characterize as a grind, but his methods were brilliant, Darwin said, and his results often amazing. Since he had been the star pupil of Darwin's secret collaborator, Darwin assured me, he had to be "the best."

Whereas Mary's job seemed primarily to help Darwin with the egg-procurement, implantation, and surrogate programs, Paul's duties centered on the body-cell problem—which was, simply put, to get a body-cell nucleus into an egg cell intact, functioning, and able to divide at the right time. Paul was an expert microscopist and cell culturist, and Darwin readily admitted that Paul knew more than he about cell anatomy and metabolism. Paul would eventually make great breakthroughs in cancer research, Darwin prophesied, and added that if everything worked out, both Darwin himself and Paul could count on Max for continued funding in their respective areas of research.

Contrary to prevailing scientific wisdom, Paul seemed to think that nuclear transfer might be successfully achieved through microsurgical manipulation as easily as through cell fusion. Even while James Watson was saying that the microsurgery employed by Gurdon in the successful cloning of frogs' eggs would prove too cumbersome for human cloning and that cell fusion was the path to follow, other researchers, Paul said, were demonstrating that human body cells *could* undergo extensive microsurgery and survive.

Human cells had long been regarded as simply too small and too complex to suffer the interference of clumsy human fingers in their internal affairs. This was no longer so, Paul said. He spoke of researchers who had been able to isolate and culture individual cells derived from both normal and malignant tissue. These researchers had been able, microsurgically, to pluck *individual* chromosome pairs from the nuclei of these cells and microinject them into other cells. The doctored cells could then be cloned to produce multiple, identical copies of themselves, demonstrating, among other things, that the microsurgery and the exchange or addition of nuclear parts had not destroyed the cells' ability to go on dividing and growing.

The researchers also demonstrated that they could inject a variety of chemicals and even virus fractions into human body-cell nuclei without destroying them. It was recognized that these microsurgical techniques "could prove of incalculable value in discovering how both normal and cancer human cells tick," Paul said, quoting from a report he had clipped. "The ability to carry out micro-operations on human cells could also bring nearer the possibility of 'clonal man'. . . ."[24]

Some other researchers, Paul continued, had recently succeeded not only in microsurgically disassembling the nuclei, cytoplasms, and membranes of cells—including living amoebae—but also in reassembling them and getting them to grow again. A living creature had thus been killed, for all practical purposes, and then brought back to life. Moreover, it was not necessary to reassemble these creatures from their original parts. The disassembled parts of many amoebae of different strains could be scrambled and viable, living creatures reconstructed. This was, when you thought about it, a mind-boggling development, one with implications more far-reaching than those of human cloning.[25]

Human cells scheduled for microsurgery by Paul were grown on glass cover slips in culture mediums. These cover slips were inverted, forming the "lid" of microsurgical chambers into which microtools were introduced through tiny holes. Low magnification was used to isolate the individual cells for each experiment. Once located, these surgical candidates were scrutinized at much higher magnification and either chemically or mechanically separated from their neighbors through the use of cell-cement solvents or microneedles. The unwanted cells would simply be scraped away and permitted to fall to the bottom of the surgical chamber. The target cell, still attached to the glass cover slip, was then operated upon.

Many different microsurgical procedures were possible. I watched Paul as he microinjected chemicals into the nuclei of some cells that he had originally scraped from the lining of his own mouth. The needle he used had a diameter of a fraction of a micrometer (less than .000039 inch). In some cases these injections would encourage the nuclei and cytoplasms of these cells to separate. A microaspirator, with a somewhat larger internal diameter, was used in another experiment to suck nuclei right out of their cells. The microinstrumentation was carefully calibrated to "deamplify" the microsurgeon's movements, and the pressures required for microinjection and microaspiration were under automated control.

Paul was convinced that, given time, and through the adjustment of several dozen chemical, environmental, and mechanical variables, he could produce a successful nuclear transfer through microsurgery. By simply altering slightly the preparation of some of his microtools, for example, he said, he had bettered his initial results very significantly. These preparations involved refinements in the microforging of the tools and in their lubrication just before inserting them into the cells. By adjusting other variables, such as the presurgical temperature of the eggs and body cells, he was confident he could make still other gains.

It developed that Darwin was somewhat distressed by Paul's preoccupation with the microsurgical approach to nuclear transfer. It was like Paul, he said, to pursue "the greater challenge." Darwin believed they should put most of their effort into cell fusion. There was a lot to learn through cell surgery, no doubt, but their task was to clone a man, and that, he felt, might be most easily and safely accomplished through those techniques that would require the least meddling.

Darwin said that what he wanted was a reliable technique whereby the appropriate body cell could be isolated from its neighboring cells and the nucleus then persuaded to part company with its cytoplasm *without* cutting, scraping, or prodding it. Then he wanted an egg that would similarly cast out its nucleus, preferably without its even having anything injected into it. Finally, as the deus ex machina in the scenario, he wanted something that would swiftly and neatly marry the two parts of his plot without altering either one in the process. All of this could be done, he believed, though some compromise might be necessary. There were still problems to be solved. But advances in cell fusion had been coming along at a rapid pace, and he was confident that he and Paul would score a few of their own.

As early as 1960, he told me, it had become possible to fuse together the body cells of unlike creatures, using a technique developed by two Parisian researchers and later refined by scientists at Oxford University in England. The technique caused something of a sensation in the scientific world when it first

155

surfaced, since it enabled man to do something nature had seemingly always forbidden: to intermingle the genetic material of two
different body cells (men and rats, chickens and yeast, and so on)
—in ways that could never be achieved through natural or selective breeding. There was much to be gained by this "breach" of
the natural law, for fusion provided one more means of studying
the functions of individual cells and, to a lesser degree, individual
chromosomes. It was, in a sense, the precursor to the recombinant
DNA work, in which the even smaller gene packages could be
moved about and thus analyzed at will through what amounted
to man-made mutations. Even now, cell fusion remained of considerable interest in the realm of chromosome mapping, the study
of membrane interaction, and the like.

The method that arose in 1960 exploited the odd and not
fully understood capacity of some viruses (particularly the influen-
zalike paramyxoviruses) to induce body cells to stick together, to
form intercellular "bridges" by which cytoplasmic material is
exchanged cell to cell, and then gradually (or, in some cases, very
rapidly) to fuse completely, forming large single cells with con-
joined nuclei. When these fused cells divided, it was observed
that the resulting "daughter" cells often contained all of the
chromosomes of each of the original single cells, indicating that
the fusion was indeed complete. The virus used in 1960 was the
Sendai virus, rendered noninfectious by exposing it to ultraviolet
light. Since that time, numerous other viruses had been demonstrated to be similarly "fusogenic"; that is, capable of causing two
different cells to merge. Some, like the Germiston virus, could
effect cell fusion with amazing rapidity—in four or five minutes
in some cases.[26]

Paul, though he did at times seem to be biting off more than
he could gracefully chew, said he had not lost sight of the goal.
The reason he was putting so much effort into microsurgical
experiments was not that it was "the greater challenge" but simply that he thought it might work better than chemical or virally
induced cell fusion. The problem with fusion was that it could not

156

be so easily controlled as microsurgery. With surgery, he said, he had, so far at least, been able to achieve greater precision in "cell-cycle synchrony"; that is, in getting egg cytoplasm and body-cell nuclei to divide at the same time. Moreover, with microsurgery it wasn't necessary to denude the egg of its outer protective layer. Nor did the egg and body cell have to be mixed together in a common medium (possibly ideal to neither) for exposure to the fusing agent. There were advantages and disadvantages in both approaches. Paul said he wanted to assess these fully before committing the project to one or the other.

This seemed sensible to me. In any event, Paul was not neglecting the fusion approach. Among other things he had been experimenting with was a group of fungal metabolites that a British researcher, S. B. Carter, had shown in the mid-1960s to have a number of curious effects on the behavior of cells. These metabolites, which had come to be known as "cytochalasins" (from Greek roots which meant "cell relaxers"), could, at some dosages, stop or slow down cell division and, at others, cause the cytoplasm of cultured cells to expel their nuclei. It was later found that filtrates from numerous other fungal species possessed similar powers, but the substance that had been most fully investigated was one called cytochalasin B.[27]

It was a fascinating substance, Paul said, not only because it could act like a "chemical scalpel" but because, by carefully controlling its concentration in the cell medium, you could use it to halt cytoplasmic division completely, while leaving nuclear division unaffected. By exploiting this characteristic, you could perhaps further synchronize eggs and body cells.

Paul showed me video tape of the cytochalasin B in action. Within minutes of exposure to the fungal agent, cell nuclei could be seen to move toward the edge of their respective cells, putting bulges in the cytoplasmic membrane walls. Soon they pushed through the walls, which closed behind them. They often remained attached to the vacated cell by a thin cytoplasmic umbilicus. Paul demonstrated that he could reverse the process, with

157

the nuclei magically—or so it seemed to me—reentering their cells, simply by putting the cells in a new medium, one free of the fungal agent. No ill effects on the cells had been noted even after long exposure to the fungus, which seemed to work by altering cell-membrane tensions, perhaps both internally and externally.

In 1970, Paul noted, cytochalasin B was successfully used to remove the nuclei of various mammalian cells. The nuclei were undamaged, and the cytoplasms of cells thus deprived of their nuclei also remained intact. Soon researchers demonstrated that cells disassembled in this way could also be joined together with the Sendai virus. One research team concluded from this work that the way might now be open to clone mammals.[28]

As far as he knew, Paul said, no one had rushed to follow up on this tantalizing suggestion. At least no one had done so on the record. The closest anyone had come in the intervening years was to use cytochalasin B to introduce a mouse body-cell nucleus into a human cancer cell.[29] This was a step bolder than any previously described in the literature, but it would be some time before we would see news of any effort aimed specifically at cloning mammals or even of any effort to use the new cytochalasin/fusion techniques for nuclear exchanges using *only* human cells.

There were a number of developments that pointed toward and, in a sense, flirted with mammalian cloning in addition to those already described,[30] but none, so far, that were aimed specifically at that end. Actually, as it turned out, there was one such effort under way, but the world was not to see anything on that in print for another year. By that time the fusion efforts of Darwin and Paul would be yielding very promising results.

CHAPTER 25

When I left the laboratory in mid-January of 1975, I flew east to California, where I met once again with Max. My flight from Hawaii, where I had stopped over, arrived in San Francisco in the evening, and Max himself picked me up at the airport. He was wearing a white turtleneck and a blue blazer and was growing a beard, which he described as a "periodic aberration." I told him it made him look even younger—and it did.

On the way to his house, we stopped at the Trident in Sausalito. It was, as Max put it, "a sort of health-food place" with a great view of the San Francisco skyline just across the bay. Over some avocado concoction, Max told me he was relieved that Darwin was cooperating, so that that now nothing was being kept from me. He regretted Darwin's earlier "paranoia." I didn't comment on Darwin's version of things.

We talked for some time about a new business venture, one that seemed to excite Max as much as the personal adventure he was experiencing in league with Darwin and the rest of us. This new undertaking seemed too ambitious even for a man of Max's means and, though I didn't say it, certainly for a man of his age. It involved, among other things, sinking vast amounts of capital into a politically unsettled country. The project might be years, decades in progress. It would require constant surveillance and crack judgment. But Max said that he felt renewed by the almost unmanageable scope of the project.

There were aspects of this project that disturbed me. There were things about Max that had emerged over these many months that sometimes also disturbed me. Still, at no time did I ever really regret my decision to assist him in a fashion that would have the effect, in a sense, of perpetuating him. Granted, there was no real telling what his clonal offspring might be like. But, or so it seemed to me, the impersonal, multinational machine that in the absence of a flesh-and-blood heir would absorb Max's wealth and power was more to be feared. An individual could at least be identified and dealt with. An impersonal corporate entity could not be so easily engaged.

There was a good chance that Max's offspring might be like so many offspring—different from their parents, capable of seeing the world in new ways. Max, I was sure, hoped for this himself. There were things in his past that he regretted. He was oppressed at times, I was convinced, by the feeling that he had been the victim of circumstances beyond his control. To some extent he had become the person others had wanted him to become, not for his good but for theirs. These circumstances and these persons had prevented him from realizing the greater, untapped potential that he believed—for he said so expressly—was within him but now seemingly unreachable. He perceived a bad side to himself that he hoped his clonal "self" could transcend or, more likely, avoid in the first place.

Max said that, even though he might not live to see the completion of the project he was launching, his "son" almost surely would. The boy might, he said, dismantle the whole thing. That didn't matter, for if he did, he would presumably have good reason for doing so. Max, he now admitted, had come to doubt that his own perspective was necessarily the correct one. Not that he had been a long time arriving at this state of doubt, he said, but he *had* been a long time acknowledging it, even to himself. The source of his self-mistrust, it seemed to me, was at base this feeling which had been his from early childhood—that he was out of harmony with his own inner destiny.

In a lighter moment, Max commented that his cloning would do more than give him a second chance; it would also make a lot of people very unhappy. No one of his corporate heirs-apparent, he noted, had any inkling of what he was about to visit upon them. In fact, they probably wouldn't know until he died. This raised a number of questions I took up with him the next day. How would he secure the child's right to his estate? What would happen if he were to die while the child was still very young

—legally a "minor"? What about the surrogate? Again in the event of Max's death, could she claim to be the child's mother and therefore his legal guardian, with all the perquisites of that station? Might she not be easily manipulated by those other parties who/which would be very interested in the distribution of Max's estate? Would Max have to adopt his own clonal offspring, even though more "him" to begin with than any "real" son? Would he have to marry the surrogate to confer legal rights to her "son"?

We had touched upon some of these things in the past, but they had seemed a bit premature then, and Max had been unwilling to discuss them in detail. Now I also wanted to know what plans he had for rearing the boy. Again, what would happen if Max were to die while the boy was still very young? Would he be educated according to a prearranged program? How could Max be confident that his wishes in this regard would be honored? Did he not fear that some might want to do his "double" harm? Either someone who possibly regarded cloning as an affront to God or nature or someone who was thwarted by the child's existence and stood to gain by his demise?

The answers to at least some of these questions emerged over a period of time. The child's right to the estate, Max said, would be established at base by the universal laws of blood relationships. The child would be, no doubt about it, the next of kin and then some. He said that, when the child was born (he, too, avoided the word "if"), he would independently verify its clonal origin with the appropriate scientific tests. Neither Darwin nor I would have any say in his selection of the individual or individuals who would perform these tests. That verification would be duly witnessed and notarized and put away for safekeeping. Darwin, Paul, Mary, and I would also be required to swear to the essential events leading to the live birth of a cloned individual. These documents would also be sealed and safely stored. They would be made known only in the event that the child's heritage was at some point seriously challenged.

I was astonished at first when Max told me that he had

already made plans to have himself interred in a cryogenic vault when he died. His frozen body would thus be preserved indefinitely. He had little hope of being thawed out and revived, he admitted, although to his way of thinking this was not impossible. Several decades down the road, there just might be breakthroughs that would enable medical science to restore a cryogenically "suspended" person by thawing him/her out and correcting the defect or damage that had led to death. The mere presence of suspended souls "gathering dust," as Max put it, in icy vaults would provide some measure of challenge to future generations of scientists, who would be tempted, he was sure, to try their hand at "bringing back the dead."

To make this even more tempting, he was leaving a special fund which he believed would be sufficient to keep him in the deep freeze for at least a hundred years and to reward the party or parties who might eventually "rescue" him. But, fearing that he might be used as a guinea pig before the time was ripe, he was stipulating that no effort be made to revive him for at least fifty years. Thereafter, a medical school he planned to designate could undertake to thaw him out, either "in house" or through "subcontract." Success in bringing him to life would be rewarded with a sizable lump sum which would, however, otherwise remain frozen as long as he did.

In a codicil to his will, he told me, he was also stipulating that in the event of his revivification he was abandoning any right to his "former assets," lest the prospect of his second coming strike terror into the hearts of his heirs, if any, and so induce plotting (of the possible plug-pulling variety) among them. He said he had planned all this even before the cloning idea had become compelling, but now it occurred to him that his frozen remains might prove useful in a practical sense: in case of a battle over his estate, tissue could be taken from his frozen body sooner than fifty years, if necessary, to be subjected to comparative analyses, used for grafts, or otherwise employed as a means of proving the clonal origin of his offspring and heir, should the documentary

162

evidence turn out to be insufficient. He would provide for this possibility as well, he said.

He hoped that the disclosure of secret documents or the deicing of his "frozen hulk" would never be necessary. He expected to live another twenty or thirty years at least, but even if he did not, he expected the new will he was already drawing up would be sufficient to protect the rights of his "copy." Potential manipulators would be heavily guarded against by various specific provisions of the will, and in the event of the heir's death prior to his reaching legal age, all assets would revert to designated charities or be frozen for specified periods of time, depending upon various circumstances. In case Max predeceased his offspring, he was also stipulating that certain persons, myself included, have guaranteed access to the heir. A complex set of rules, I gathered, was being formulated to govern the guardianship of the child should he be prematurely "orphaned."

Little if anything was being left to chance. He felt confident that, if he were to die even while the child was still very young, the transfer of power could take place without revealing the infant's clonal beginning. To help guard against the sort of challenge that might make full revelation necessary, he hinted that he might marry the surrogate or adopt her or the child or both. If he did marry the surrogate, he might draw up a separate contract with her, so that her rights could never overshadow or absorb those of the child. This might prove useful, he added, whether he married her or not.

To my questions about possible fears for the well-being of the boy, Max conceded readily that he had some. For that matter, he feared for his own life from time to time. Engaged in as many enterprises as he was, he had inevitably made his share of enemies. He was sure there were people who, if they knew what he was up to, would lose little time "sharpening the long knives." Max was a man who was expected to die without a real heir.

Max said he intended to give his clonal heir money *and* control. A complex legal instrument with its own checks and

163

balances was being "engineered" to help the heir—in infancy, childhood, adolescence, and early adulthood—retain control of Max's interests and assets. Max had anticipated and identified every possible source of trouble, he said; the documents that would govern his estate would keep all possible usurpers "divided and at each other's throats" for years to come, while incidentally "boosting their productivity."

In addition to all this, he indicated, a contract might be drawn with Darwin whereby, in the event of both his and his clone's deaths, cells might be removed (probably from the clone) and used to produce a "replacement beneficiary," with Darwin himself, in this instance, serving as the executor of the estate for a limited period of time. This raised another question that had been nagging at me. Did Max have any intention of stipulating that his clone, in order to receive the estate, agree to have *himself*, in due course, cloned? I could imagine Max at least thus symbolically perpetuating himself forever. He denied any such intention, but added that he could not prevent his clonal offspring from opting for a clonal reproduction of his own. On the other hand, he would not have any objections or even be disappointed if the clone decided to marry, have children in the normal fashion, and otherwise lead a biologically conventional life.

It was an "experiment," Max insisted; he would be interested "whatever happens." He reminded me that he would not try to re-create the environment out of which he himself had emerged—that was the last thing he wanted. The child's education would be of the highest standard; Max was taking care to ensure that, whether he lived or died. Most important, the child would have the security of a firm identity, of knowing precisely where he had come from. Max did not believe that this sense of knowing might become overpowering. His offspring, he insisted, would not be a "satellite," but rather "a new planet." After all, he would be as different in one way from Max as from the rest of mankind—the only human creature so conceived. Yet that conception, while unique, would not, he felt, cause anxiety or fear once everything was explained evenly, rationally.

Still, wasn't there the chance that his offspring might come eventually to regard him as an arch egotist, nonetheless? So might any child, Max countered. Reproduction by any means encompassed some measure of ego. As far as he was concerned, we were merely splitting hairs. Anyway, he hoped to love this child not as he loved himself, for that he could not do even if he wanted to; no, he would love his "son" as he would love someone he "wanted to be."

CHAPTER 26

The girl was "going on seventeen," as Max put it. She had been orphaned when she was ten. Her parents and two of her siblings had died violent deaths, unlucky bystanders in what he called a "bush-war skirmish." The girl's right hand had been badly burned in her escape from the blaze that consumed her family's hut. The hand was now withered and deformed, and of only marginal use.

Max showed me pictures and a short video tape of her in February, when we again met in California. She was not smiling. She was very pretty, but she seemed unlike most of the girls and women I had seen in the region during my visit. She came from that area, Max explained, but had a mixed heritage. He said she was very mature for her years.

I asked him how he had found her, thinking perhaps he had seen her in one of the orphanages his money helped support. I knew he had been surveying prospects in those places as well as in the factories and on the plantations. I thought from the beginning that an orphan surrogate would appeal to him and of course, from a logistical point of view, simplify things considerably. No, he said, the girl, whom he called "Sparrow" from the beginning of his account, had turned up on one of the farms doing menial labor—with, he claimed, twice the productivity of the others.

It all struck me as more than a little ironical. From what I knew of Max's background, the "skirmish" that had orphaned Sparrow and left her without the use of her right hand might well have been one that had ultimately benefited or had been at least partly provoked by men like Max. First the skirmishes, then the factories, mills, and other activities—which nevertheless, Max would insist, provided better livings for the locals than they had ever had before. Against the still-pristine view of the country Mary had painted, Max developed a picture of corruption and decadence and said that many girls like Sparrow, without his factories or farms, would starve or be forced into prostitution.

There were not nearly enough jobs as it was, he pointed out, and he would provide more, despite the politically inspired resistance he met to every move aimed at expansion and a "better life." Girls as young as eleven, he claimed, were being shipped off to cities where they staffed brothels and massage parlors. It was a complex irony, I thought as I reflected on Max's own image of himself, that could give us creditable orphanages to house the offspring of discredited wars and of men whose own greed or particular vision of the future had made their philanthropy not only welcome but necessary. At the root of all these ironies, I decided, was not so much bad conscience as a bad concept of progress.

And so, in any event, Sparrow had come "home"—to shelter under Max's protective wing. I had difficulty grasping it, in the way I have just explained, and yet no difficulty at all—for, as I read the translated transcript of one of Sparrow's interviews, I learned from Mary's analysis of the girl that Sparrow had "a strong drive to succeed . . . she is almost fierce in her resolve to control her life and reverse the bad luck that has orphaned and crippled her. . . ." The classic overcompensator of Western society, I thought, something like Max—in fact, they were birds of a feather.

Max similarly remarked on Sparrow's "strong will" and "independence." He related an incident in which Mary, with good intentions, presented the girl with a pretty glove she might wear over her deformed right hand. Sparrow had put it on, looked at it tentatively, and then had quickly taken it off and given it back to Mary with polite thanks. It would look odd, Sparrow told the older woman, to go about with only one glove. Mary had been flabbergasted and was not sure whether the girl was subtly reprimanding her or was simply naïve. There was no doubt in Max's mind as to which it was.

Sparrow had turned up as one of Roberto's recruits when Mary, tired of interviewing "children," had counseled Roberto to look for more serious or mature types. Roberto thought immediately of Sparrow, whom he had seen on previous occasions. He had always, however, rejected her because of the hand. It obviously repelled him, and he thought it would repel Max, too. His complaint that the only serious ones were "ugly" brought a sharp reprimand from Mary and renewed insistence that he start looking for brains as well as beauty.

The only compromise he could immediately come up with was Sparrow. He presented her to Mary with a defiant, I-told-you-so air, but Mary was almost instantly charmed. "This one didn't giggle," Darwin explained to me when I saw him again. Mary supervised the picture-taking herself for the first time—not, as she told me later, to minimize the impact of the hand but to make sure that Sparrow's other assets got at least equal play. Max, Mary reported with satisfaction, was also "instantly" taken by Sparrow and had never commented on her hand, though its deformity was visible in the pictures he saw and was accounted for in the report he read on her before they actually met in March.

Max was also very taken with the girl's response to the request, by then standard in these interviews, that she tell a lie. The girl had simply stated, apparently without hesitation, "I am dead." Most of the girls either giggled in response, were uncomprehending, or came up with the cultural equivalent of "I'm the Queen of Sheba." But Sparrow, Max repeated, had immediately and unsmilingly said, "I am dead."

It was possible to read too much into this sort of thing, Max knew, but to him this response was very significant; to him it indicated that to be alive was, to Sparrow, a matter of utmost, conscious concern. Most people took being alive for granted. Unlike her parents and two brothers, she had escaped death, and it was her immediate mission in life, Max said, to go on escaping death. She was a "survivor" like himself, as Max saw her, and though she might feel death was after her, she also believed she could cheat it, escape, and live—perhaps forever. He spoke with real pride of the girl whose dark eyes looked out at me so fearlessly, even challengingly, from the picture I held in my hand.

Sparrow, who had surfaced in early 1975, was soon joined by "AnnaBelle," who also arrived by the Roberto-recruited route. Max gave her this nickname because her own name was so difficult to pronounce unless you knew the language. AnnaBelle was about Sparrow's age and also very pretty, though perhaps "cute" better suits her. She was warm, outgoing, and, according to Mary, "frivolous," though in a more charitable mood she conceded that AnnaBelle was "very sweet." Mary clearly favored Sparrow, who in her brooding manner certainly presented quite a contrast. There was fire in Sparrow, too, Max said, but it was "all banked within." Sparrow was all mystery, AnnaBelle an open book.

Max liked AnnaBelle's looks, no doubt about that. But there was more to it. He also liked her energy. It was easy to imagine her a mother, despite her childlike qualities. She was more robust than Sparrow. I wondered if her place in the active file signified some second thought about Sparrow on Max's part, but he said no, that Darwin insisted upon several acceptable surrogates, since one or more might be out of phase, in terms of menstrual cycle, when needed. Darwin kept at Max to find others.

I was permitted to read Mary's dossiers on the two girls. Sparrow, it seemed, had been earning her keep and then some, practically from the day she was orphaned. An aunt who worked on one of the farms had taken her in and promptly put her to work in the fields. The girl had quickly proved that she could handle the job with only one good hand. In Mary's estimation, the aunt had been a good influence on the girl; she was a disciplinarian, a convert to Christianity, and believed in "getting ahead."

As a child, the aunt had lived in a missionary school and then had later worked there for a time, and could read and speak a European language. It was of interest to Mary that, while this woman had never taught her own children this language, she did teach it to Sparrow. Sparrow had read a translation of the Bible in this language many times over. Sparrow had a disconcerting way of using both her native and her learned language—sometimes in the same sentence—and for good measure she would occasionally throw in an English word or phrase. Even before Max began having her tutored in English, she showed an amazing capability of absorbing it simply by being around those few who spoke it at the hospital.

Mary was curious, Max told me, to know to what extent the Bible, up until then the girl's only reading material, had shaped her outlook on life. Had Sparrow taken everything she'd read literally? Apparently not; Mary gathered that Sparrow and her aunt had argued over this issue more than once—until Sparrow began humoring the aunt to keep the peace. Mary was impressed, as was I. You would have expected so young a girl to be overwhelmed by this book and the new view of the world that it disclosed. Sparrow, however, had remained faithful to the religion of her parents and her people. She found the Bible "wonderful," but its stories, she felt, were like those her grandfather had told her before he died—instructive myths.

Max had promptly provided Sparrow with a crateload of philosophies and fictions, both serious and childish, written in the language the aunt had taught her. Mary observed her consume the whole batch, frequently consulting the enormous English-/other-language dictionary that Max had also provided—so that her English improved at the same time. It wasn't long until Sparrow was using words like "absurd" and "fantastic."

Darwin later told me that Mary was a bit distraught by some of the racy titles Max had included in his literary "Care" package.

But Sparrow, for whom the facts of life were no less familiar than what Max once called the facts of death, was apparently more puzzled than titillated by much of what she read. Mary noted with some satisfaction that Sparrow viewed most of the sexual machinations of Western society as peculiar. "Do they not work?" she asked earnestly after finishing one novel in which the characters occupied their time with nothing but a mildly satirical series of bedroom romps. And what, she wanted to know, was the utility of silk stockings and corsets? And why did the women always behave "like children around the men" but then "talk grown-up" amongst themselves?

Sparrow had been taught that a girl should remain a virgin until she married, yet she had difficulty comprehending the guilty breast-beating that accompanied the loss of "virtue" that over-punctuated some of the novels she read. Virtue, in her world, had little to do with it. If you had sex with a man you were not married to, you might have his baby—and that would be a hardship. You had to live, and with children you needed a husband, someone to work while you were burdened with pregnancy or if the children were sick.

Sparrow told Mary that she believed in love, but it seemed that her concept of this oft-discussed state of being was more utilitarian than romantic. You "loved" a man who "loved" you —who would work hard alongside you throughout life. What made a man handsome or attractive to Sparrow? Strength, it seemed, but her idea of strength went far beyond physical prowess. A strong man was one in control of his own life and to some extent the lives of those around him. Mary didn't emphasize this, but it seemed that in some of the books Sparrow read, the characters she found "strongest" and thus presumably most attractive were not the good guys—who might be rather bland—but sometimes (at least by prevailing standards) the bad guys. Perhaps this was why Darwin would later on occasionally refer to her, when Max wasn't present, as "Lady Macbeth."

The unhealthiest aspect of Sparrow's view of things, I

thought, had to do with her attitude toward her own father: he had not been a "strong" man—primarily because he had not been able to protect his wife and children. He had been trampled underfoot by stronger men, one of whom had raped one of Sparrow's older sisters. Did she resent the "oppressors"? Yes, but as individuals and not as a class, and her resentment came through to me as an almost obsessive desire to be even stronger, which almost had to mean—whether or not she realized it—even more oppressive. I never discussed such thoughts with Max.

Sparrow wanted to have children because this was what women were in part designed for; having children was necessary to keep the world turning, but she hoped not to have as many children as her mother and aunt had. Two or three, she thought, would be adequate; she would gladly use contraceptives of the sort Mary told her about. It was important to her, though, to have some children. To do so would prove again that she was alive and vital, and she had a better than vague concept of perpetuating herself through her children. She hoped to have very strong, beautiful children, she said—another reason to wait for a strong husband.

Some men, it seemed, had assumed that Sparrow would go to bed with them "because of the hand," as she so knowingly put it. Her defense against this sort of contemptuousness was to be even more contemptuous in return. She had had to fight more than once to establish that she was proud of herself and would not be easily had. She repelled one amorous aggressor with a sharp tool, leaving her mark on him for life. Through this act more than anything else she had earned the respect of many of those who worked around her. Some of the men, Roberto later told me, preferred to think she was a little crazy, and that alone, in this culture as in some others, was something to respect.

As for AnnaBelle—the romantic—her view of things was no doubt far more unreal than the pragmatic Sparrow's, but at least these were the unrealities with which we are so familiar. With her comic-book level of literacy, her favorite fantasy centered upon a rich American—preferably a movie star. She came from a more affluent family than did Sparrow; it was still a very poor family, but she was more attuned to the cultural vibrations of the times. She had had access to movie magazines, spoke a little English, and talked dreamily of various singers and other pop stars.

AnnaBelle's sister had married a petty bureacrat, "rich" enough to own a phonograph and rent a "modern" house with running water and flush toilet. They also owned a television, though as yet there was no programming in the area. AnnaBelle lived in breathless anticipation of her next birthday, for on that date, she told Mary, her sister had promised to give her some eye shadow and other makeup. She hoped she could go back to the factory where she had worked to show the other girls.

In Mary's view, AnnaBelle's virginity was not likely to endure much longer. In her conversation with Mary, the girl spoke often of "feelings" she had around boys, but when pressed to explain further would only dissolve into giggles. When Mary, trying to get at the truth, suggested that perhaps she had already succumbed to those feelings, the girl appeared almost frightened. No, she hadn't, she insisted. Her sister, apparently, had warned her time and again to play "hard to get" or she would end up with something less than a petty bureaucrat for a husband. Mary thought it was now a matter of waiting to see which would prevail —the "feelings" or visions of flush toilets and television. Because AnnaBelle was not cunning or patient, Mary was betting on the "feelings"—unless the girl was essentially isolated from such temptations.

And so "Lolita," as Darwin dubbed this one, was given a private room at the hospital, complete with stereo, picture books, and eventually a video-cassette machine. Her sister was delighted by the arrangement and told AnnaBelle to behave; she was not likely to be selected to take part in such a study again. The sister was told that Darwin was doing research on various fertility factors and that this research might eventually make possible the transfer of an embryo from one woman to another. Thus women with damaged tubes or wombs could, through the use of a surrogate, have babies that were genetically their own. No one said that AnnaBelle would be such a surrogate, but the sister apparently indicated that she would have no objections if she were.

Sparrow, on the other hand, was treated differently. She was told at the outset that the study involved a couple who hoped to have a child through the use of a surrogate. She was asked to keep this information confidential. The books, the money, the tutoring in English—all of these were gifts to Sparrow from this couple. Sparrow was quick to grasp the whole thing. It did not shock her. And the uniqueness of her role in the scenario was not lost upon her—such things, she said, had been in a sense foretold in the Bible. She spoke of the Virgin Mary and the immaculate conception.

CHAPTER 27

My second visit to the facility took place in the early summer of 1975. Max and I went there together, aboard one of his planes. We were excited because Darwin had informed us that he and the others were beginning to get what he termed "good development" with some of the nuclear transfers—body-cell nuclei implanted in enucleated eggs (eggs chemically or surgically deprived of their own nuclei, as mentioned earlier). With practice, patience, and continued fine tuning of the culture media, they had come by ever better results. It wouldn't be long, Darwin felt, until they were ready to make their first cloning attempt.

When we arrived at the facility, though, we found Darwin in a particularly agitated frame of mind. An incident had just occurred that would subsequently force Max to give Darwin a freer hand with the surrogates. Mary, Darwin, Paul, Max, and I all gathered in the lab to talk about the event. The door was carefully locked. What had happened was this: one of the nuclear transfers unexpectedly advanced in culture to the morula (32-cell) stage—that fully energized stage at which it might successfully be implanted. This was the first time this had happened. Even the "good development" that Darwin had reported before we came referred to transfers that had advanced no further than the 8-cell stage. Darwin, Paul, and Mary watched with fascination as this embryo grew, expecting it to stop dividing and disintegrate at almost any moment.

When it finally achieved the morula stage, they realized that neither AnnaBelle nor Sparrow was in the proper phase of the menstrual cycle for implantation to succeed. Nor was there time to try to induce receptivity artificially through hormone injections. Darwin was opposed in any case to tampering any more than was absolutely necessary with the surrogates' biochemistry; you had to keep everything on as even and natural a keel as possible, Darwin believed, or you could not maintain a pregnancy.

Mary was explaining all this when Darwin, who was fidgeting and sweating with apparent nervousness, cut in to say that they had tried to reach us by telephone but we were already on our way. An unplanned stop had further delayed our arrival and lengthened the period in which we were out of touch.

"We had no choice but to go ahead," Darwin blurted.

It took us a moment to figure out what he was saying. Then I thought, Good Lord, they've implanted it—and not in a surrogate selected by Max. But if Max was dismayed, he did not betray his emotions; he simply asked for a fuller explanation.

Mary said that a frantic search had been launched to find a woman in whom the embryo could be implanted. They decided that, since they could not reach Max to get his view of things, they should proceed; the chances were that he would say yes, in any event. They couldn't wait because the egg might stop dividing at almost any time—unless they were to get it out of the test tube and into a real womb. If a pregnancy resulted and Max was

174

adamantly opposed to having his clone borne by an impromptu surrogate, an abortion could always be performed. There had been no time to waste.

The logical choice for the surrogate, Mary continued, was the woman who had donated the egg in the first place, inasmuch as her cycle would be in the proper phase to take it back again. But there were problems. She was a married woman, for one thing. She had agreed to give up some of her eggs in return for having her tubes blocked as a birth-control measure. Still, she would do—if she could be found. She seemed to have taken the money she was paid for her troubles and gone off on her own; her husband was almost as upset by her absence as Darwin was.

Subsequently, nearly every woman at the hospital of child-bearing age, starting with the nurses and aides, was queried as to menstrual-cycle status. Finally a patient was located who had been using the rhythm method of birth control and, at least by her chart, was in about the right place at the right time. She was told that "medical science" needed her. She agreed to undergo some "tests" and participate in a "study." She would be well paid, she was promised. The plan, in the panic of the moment, was to act now, explain later—if necessary. This woman, too, was married. Her family, she was assured, would be told that she was being kept at the hospital for a few more days but that everything was all right. The relatively minor condition that had brought her to the facility in the first place had nothing to do with her having to stay on a bit longer, she was assured. She was "a little frightened" but malleable. The implant, Mary concluded, had just been performed.

Darwin admitted that he was too rattled even to think about what should be done next—if, indeed, the woman did become pregnant. The woman had not been prepared for the possibility of a test-tube pregnancy and, Mary felt, might well prove unable to grasp or cope with the idea. On the other hand, no doubt she could be persuaded to believe easily enough that she had miscalculated and become pregnant in the normal fashion—unless by some fluke she and her husband had not had relations in some days. But then the baby would have to be acquired from her somehow after it was born. Perhaps she would sell it; perhaps, Mary said, shocking me at least, they would simply have to kidnap it. Darwin nodded.

Max still had said practically nothing, and Darwin again leaped in. This might all have been "incredible luck," he said; it was possible they would never get this far again. They couldn't afford *not* to take advantage of the situation. Max said only that he wanted to take a look. And so the whole lot of us trooped into the room where the woman was in bed. She looked considerably more than just a little frightened. Mary told her that we were visiting doctors touring the hospital. She asked the woman if she was comfortable and took her blood pressure while the rest of us looked on. The woman was no beauty, and none of us had any illusions that Max would take any personal interest in her.

"Well, what could we do?" Darwin said, once we were out of the room. We were "very, very lucky" to have got this far, he said again. Max said he was pleased and that of course they should proceed, but his disappointment was evident from the flat tone of his voice. Clearly, this did not fit into his plan.

As we all knew, of course, there was a very good chance that the embryo would not take hold, that it would not attach itself to the lining of the womb, and that no pregnancy would occur. And in the end, this was what happened. The woman at no time showed any sign that she was pregnant and in due course was dismissed from the hospital.

Max was persuaded by this mishap that Darwin must have more surrogates, as he had been insisting all along. Max went back over the files and selected two more girls from those he had previously rejected. Both were summoned at once, and both were delighted to have new jobs at the hospital and away from the tedium of the factories in which they had worked. Neither girl was Max's type; that is, neither was sufficiently endowed with those physical and mental traits that would make her irresistible to him. Still, both were virgins, pretty, healthy, and reasonably intelligent. They would do.

Both of these surrogates were told that the work that was going on could lead to embryo transfer, but it was never explicitly stated that they would be surrogates in such a procedure. Mary was satisfied, however, that if either of them became pregnant she would be able to accept her role. Darwin had decided that it was too risky to make even embryo transfer, let alone cloning, appear to be the real goal of his work. It was not inconceivable that word might leak out and attention be focused upon him.

Now that the panic of the morula incident had subsided, Darwin described his progress to date not as "incredible luck" but as the product of "systematic research" and of "the law of averages." They were beginning to get "hits"—nuclear transfers that showed signs of being able to divide and develop normally—in part because they were trying more often, in part because they were mastering the techniques required to handle and synchronize egg cells and body cells. And they had made what Darwin termed some "biochemical breakthroughs" that he said further encouraged body cells to "forget their specialties and start all over again."

On this matter of trying more often, I brought up the work of Dr. Gurdon and his cloned frogs. Despite his hundreds of attempts at cloning, I said, he had achieved a success rate of only 1 or 2 percent. I asked Darwin if he thought he would have to make hundreds of attempts before he could reasonably expect to clone a man.

The matter came up while we were all sitting around the dinner table at the house we shared with Mary during our visit. As Darwin answered me, he stabbed the air from time to time with his fork. It was "comforting," he said, to many lay people and also to some scientists to imagine that cloning was a very difficult thing to do. These doubters always pointed to experiments which were never specifically designed to clone. No one had ever gone "straight for the throat," he said emphatically; *no one had ever tried to clone a man*—or even a mammal, as far as the literature up to that time revealed.

As for Gurdon's enjoying so little success, that was untrue. If you looked carefully at the Gurdon work, you would discover that the Oxford researcher had actually done very well. Darwin cited one of the major Gurdon experiments in which fully 25 percent of the eggs in which he had injected body-cell nuclei developed in such a way that a substantial portion of them might reasonably have been expected to give rise to healthy tadpoles, had they all continued to be used in experiment. Because of Gurdon's limited time only a small number of these were selected for the full cloning treatment, but of those that *were* given the full treatment, 43 percent resulted in viable tadpoles. Another Gurdon experiment had yielded a 55 percent success rate when the complete process was applied. The 1 or 2 percent success rate so often cited held, Darwin said, only when you considered all the eggs Gurdon worked with, including the large majority that never got the full treatment.[31]

Apart from simply trying harder—or more often—what, Max wanted to know, was contributing to the better recent results? Darwin said he had found that when he added an additional major step to the cloning process he could obtain better development. Sometimes, after a body cell nucleus was implanted into an enucleated egg, the resulting embryo would divide once or twice or three times and then "poop out," as he put it—simply stop dividing and die. But if you were to take just one cell from one of those 4-cell or 8-cell nuclear-transfer embryos before it reached the "poop-out" stage and use that one cell's nucleus to create a new embryo (by implanting it into a freshly enucleated egg) you might get "a livelier performance" the second time around.

The reasons for this were not clear; perhaps the stimulation of initial cell division followed by reexposure to the egg's complex cytoplasmic chemistry promoted the better development. Gurdon had also had his greatest success with serial transfers, but for his fresh starts he had used nuclei taken from cloned embryos that

178

had already reached the blastocyst stage. But these were imperfectly developed blastocysts in which the individual nuclei had enjoyed twice as much time, owing to egg abnormality, in which to replicate their chromosomal material (DNA) before undergoing their first division in the egg. As a result they were far less likely to get torn apart by egg division and thus far *more* likely to promote normal growth.

Gurdon's serial transfers worked because they synchronized eggs and body cells and perhaps because, as Darwin thought, they further relaxed the body-cell nuclei and promoted dedifferentiation—that is, loss of specialization (as skin cells, liver cells, and so forth), so that all of the genes within the cell could be expressed and a whole new individual created.

In addition to this, Darwin said, he and Paul were having good luck with cell-cycle synchrony—culturing body cells in such a way that most of them would be ready to undergo cell division at the time they would be fused with eggs. Again, the idea was to prevent their being ripped asunder before they had replicated their DNA. Darwin and Paul were achieving this through both chemical and atmospheric adjustment of the cell cultures and incubation environments. And here again, with the refinement of these techniques, they were getting better results.

While Paul had continued to experiment with microinjection, cell fusion was the method that resulted in the 32-cell embryo they had just implanted. They had become expert in the handling and "conditioning" of "karyoplasts." These were the body-cell nuclei encased in thin envelopes of cytoplasm. It was important that just the right amount of cytoplasm be retained. Too much and you interfered with the machinery of the egg; too little and you killed or impaired the functioning of the nucleus.

Darwin said that the makeup of the medium in which the nuclei were briefly deposited prior to swift fusion with egg cytoplasm was of utmost importance. Paul and he had made some discoveries related to some of the proteins and enzymes contained in eggs and embryos. These, too, had furthered the work and

179

might prove very useful in work unrelated to cloning as well.

Darwin, who by this time had enjoyed at least three glasses of wine, said that in his opinion no one would match his accomplishments for another ten years, at least. Then, embarrassed a bit by his own immodesty, he added that this would be so partly because others would be afraid to try.

As a matter of fact, however, before the year was out we would learn of the work of an Oxford scientist who had gone, if not "straight for the throat," then in at least an only slightly wavering line. This researcher reported in *Nature* that he had activated rabbit eggs with cold shock, used Sendai virus to fuse them with rabbit body cells, and had achieved, out of numerous efforts, four embryos that divided regularly at normal rates all the way to the morula stage, at which point they might conceivably have been successfully implanted, had the researcher been prepared to go that far.[32]

Some time after that, there appeared another report of which we took special note. A research team had used cytochalasin B, the substance that could make cells give up their nuclei without damaging them, to enucleate *human* cells. In fact, these researchers were the first to demonstrate on the record that you could put the nucleus of one human cell into the cytoplasm of another human cell (the original nucleus of which had been removed) and produce new cells that thrived and grew in culture at the same rate as undisturbed cells.[33] The cytoplasms in this case were those of body cells, but there was no reason they couldn't have been the cytoplasms of egg cells. The rest of the world, Darwin had to acknowledge, was catching up, even if most of the world didn't know about it.

CHAPTER 28

During the remainder of this visit I watched experiments, and Max watched the surrogates. Now that there were four surrogates, Darwin felt that he would be able to make rapid progress. It was his plan, he told me, to implant an embryo in each of them every month if possible until, as he phrased it, "it happens." In some cases he would take the eggs from the surrogates themselves; in other cases he would get them from patients and other recruits. An effort was already under way to match egg donors with the surrogates; that is, to find women whose menstrual cycles were similar to those of the various surrogates.

As he always did, Darwin took advantage of Max's presence to obtain more body cells. Max, I knew, had been making frequent trips to the facility, partly in order to keep Darwin supplied with fresh cells. The doctor had found that he got better results when cells were not grown in culture for too long; on the other hand, some of them could not be cultured for too short a period either. It all depended upon the individual type of cell.

Darwin often commented that Max, despite his advanced years, had "young cells." Darwin had studied some cellular aspects of aging and said that Max's cells showed certain biochemical signs of being those of a much younger man. Max attributed this to his diet. Darwin was certainly no advocate of vegetarianism, but he did say that Max's nearly lifelong avoidance of fats and sugars had no doubt benefited him; and over a period of time Darwin became quite interested in some of the supplements Max took regularly, including such things as various trace minerals, ginseng, and large doses of vitamins A, C, and E.

Darwin scoffed at ginseng at first; later, when he heard that his hero Henry Kissinger was a regular user of the stuff, he looked into it and became quite excited over the papers of some respected Soviet scientists. These scientists seemed to believe that ginseng has a noteworthy antiaging effect.[34] Similarly, Darwin stopped referring to Linus Pauling, the Nobel Prize winner and vitamin C advocate, as "a genius who has strayed outside his field of competence" when he heard that some United States government researchers had demonstrated that vitamin C, apart from being a potent antioxidant and thus a cell life preserver, has the ability to increase significantly the production of white blood cells, the body's front-line defenders against attack by all nature of infectious invaders.[35] Anything that could make cells grow better and faster and stay younger longer had to be regarded by Darwin, in his present effort, as a friend.

Darwin had a variety of techniques for obtaining body cells. There were special instruments available for performing different sorts of biopsies. When Darwin obtained liver cells from Max, for example, he used a special needle (pushed through the skin below the ribs) to pluck up a sample. He could have attempted to implant the nuclei of these cells immediately without bothering to culture them, but experience had shown that better results could be obtained if the cells were permitted to proliferate in a growth culture for a period of time. Then, instead of using any of the original cells, he concentrated on their offspring. Cells that grew out in culture seemed less specialized than their parent cells and thus more suitable for nuclear transfer.

For obtaining skin samples, there were instruments for scraping and cutting. Some of these instruments had small tubular blades with which one could lift out cylindrical sections of skin. Even liver biopsies were performed with only local anesthesia—which is not to say, however, that they were without risk. Max told me that Darwin had merrily informed him he had a 1-in-5,000 chance of dying as a result of having his liver needled.

I watched as Darwin scraped and punched out little bits of skin from various parts of Max's body, including the inside of his mouth and the top of his nose. These bits of tissue were grown in culture mediums on glass discs that were designed to be fitted into centrifuge tubes. The idea was to get new cells that grew from the original "explant" of skin or tissue. Once they were present in sufficient numbers, Darwin stuck the cylinders on which they had grown into the centrifuge tubes and spun them in a solution that contained cytochalasin B, among other things. This chemical plus the spinning action made the cells give up their nuclei, which then collected at the bottom of the tube in yet another solution, especially prepared to protect them and prime them for fusion.

Eggs were similarly enucleated and then exposed at low temperature to a solution containing an enzyme that caused their outer layer to dissolve. It was necessary to get rid of this outer layer, or fusion with body-cell nuclei could not occur. And it was necessary to keep the eggs in a cold solution during this operation, or they would perish. Next Darwin put these denuded eggs into yet another medium, this one containing a virus that had been inactivated by radiation so that it was no longer infectious. The virus would cause the egg cytoplasms or "whites" to fuse with the body-cell nuclei. The next step, then, was to add the body-cell nuclei to the solution containing the eggs and the virus.

Over the next several hours the temperature of this solution was gradually elevated. Darwin added things to it from time to time. Later under the microscope he examined the cells and let me see some that had fused. Each of those little spheres, I suddenly realized, was already potentially another Max. Each of those nuclei contained a complete and precise blueprint—the architectural plans for making a man. Every gene in his body was contained therein—waiting to be unlocked by the mysterious chemical keys in the egg's cytoplasm. Every detail of his body—the color of his eyes, the shape of his nose, his sex—was encoded therein. More than that, each of those unassuming microscopic specks contained the complex genetic schemata of his brain, his mind, and, I assumed, his soul.

Well, perhaps not his *soul.* It was not the sort of thing I talked about, but I *did* think about it, this matter of soul. Perhaps that was the one part of man you could never clone. Perhaps, then, if one wanted to consider things in a spiritual sense, the uniqueness of each individual would thus always be preserved, even in a *world* of clones.

I must say that I felt vaguely uneasy, perhaps a bit awed, in the presence of Darwin's incubating embryos. I mentioned this to him, and my comment seemed to make *him* uneasy in turn. Perhaps he thought I was going to raise the bench embryo–abortion issue again. Darwin did not even like calling his test-tube creations embryos. He preferred "preimplantation zygotes."

In any event, he said, it was not a good idea to confer too high a value upon these grains of "potential life." He reminded me that the scientist's ability to keep what once again I thought of as "bits and pieces" of us alive in tissue culture was not new. It had been possible for the better part of a century.

This "awe" business, he went on, could get out of hand. Nobel Prize winner Alexis Carrel, back in the early part of the century, had been so overwhelmed by his ability to keep a chunk of chicken heart alive in a test tube (for a full thirty-three years, as it turned out) that he began to regard this homely piece of meat as holy. In fact, he eventually required his minions to don flowing, black, hooded robes before venturing into the presence of the mighty and, he thought, immortal chicken heart. The thing died shortly after Carrel did—"of a broken heart," Darwin quipped. The only tissue that was really immortal, that would apparently live forever in tissue culture, he concluded grimly, was cancer. I could be awed by that if I liked, he said.

Max, meanwhile, though perhaps not awed, was at least fascinated by the surrogates—especially Sparrow. Several trips earlier, he had met both Sparrow and AnnaBelle face to face. This had been easily arranged. Mary simply introduced him as a psychologist who was participating in the study of which the girls were supposedly a part. Max, who spoke the local language, required no interpreter and could converse comfortably with the girls.

AnnaBelle, Max said, still had no clear idea that in the course of this study she might become pregnant. Max never spoke of the possibility with her. From what he and the others said, and from what I observed, it was evident that AnnaBelle had developed a crush on Max, despite the fact that he was easily old enough to be her father, even her grandfather.

While Max's feelings, I think, were not exactly those of a father for a daughter, he was scrupulous about avoiding sexual play with AnnaBelle. There were practical as well as symbolic reasons, as discussed earlier, for maintaining surrogate virginity. Besides, Max knew that the rest of us were watching him with great interest.

Max had delighted in AnnaBelle's exuberance, innocence, and energy, but AnnaBelle was very much a child and one who, it developed, bored easily. She had difficulty amusing herself and, over a period of time, turned into what can only be characterized as the classic spoiled brat. Max already, it seemed, had begun to lose interest, perhaps overlooking the fact that what he increasingly disliked in AnnaBelle were characteristics that he had contributed to significantly through his sudden largesse.

There was another and considerably more potent factor, however, that worked to AnnaBelle's disadvantage—and that was Max's ever-increasing interest in Sparrow. Sparrow's reserve only heightened his curiosity. With Sparrow, at least, Max could talk directly about the baby. He questioned her often about her attitudes as a surrogate. If Darwin succeeded, he asked her, would she have trouble giving up the baby? She answered simply that she *would* give it up. What would she "get out of this," then? She would be paid, she said, far more than she could earn otherwise, and it would be a unique experience. She might even have done it for nothing, she said, but quickly added that she was happy to be paid. And perhaps this couple who had hired her, she suggested, might have some future plans for her. What made her think that? Well, if they had no plans for her, then why, she asked, were they bothering to have her taught English, giving her books, and providing her with information about America?

185

Sparrow was careful in the way that she posed these questions, so that she would not appear to be demanding answers. She was careful never to seem to want anything too badly, Max told me. He asked her what she thought this couple might have in mind for her. She shrugged and was silent for a while, then said that, since the wife was unable to have her own baby, perhaps she would need help raising it as well. Was that something she would like to do? She would do it, Sparrow responded in that peculiar way of hers. Was she happy at the hospital? She answered merely that it was good to be able to read books. Sparrow had learned, Max said, never to wear her heart on her sleeve, where anyone could pluck it off. A new world was rapidly unfolding for her, and to him it seemed clear that she would not want to give it up. Her seeming indifference, he added, was "as studied as a cat's."

Sparrow did not blush around Max the way AnnaBelle did. She was, if anything, the opposite of the flirt. Max said it was easy to forget that she was a child, and I agreed. He felt there was something "almost oracular" about her—that if he asked her what he called "the big questions," whatever they might be, she would come up with the answers—if only to tell him that there were no answers.

It seemed to me on the other hand that Sparrow could not but be impressed by Max, given the aura of command that accompanied him and the considerable erudition that he wielded so well. But the first real hint that she was indeed interested came during this visit, when he found that she was working on a sketch of him. She had unmistakable artistic talent, and Max had done all he could to encourage her, giving her paints, canvases, charcoal, paper, and inks. He had found the sketch by taking the liberty of leafing through one of her sketchbooks while she was occupied with something else, and she caught him at it. Neither spoke, but Sparrow looked at him a bit reproachfully, he thought. He was very pleased by the sketch, and even more by what it possibly portended.

CHAPTER 29

The first pregnancy occurred in October of 1975. I was living in San Francisco at this time, recovering from an illness that had confined me to bed for nearly a month. When Max called from Marin County to give me the news and invite me to return to the facility with him, I still felt too exhausted to accompany him. In addition, I had several free-lance assignments pressing and very little time to complete my application for a fellowship I wanted. I disappointed myself far more than I did Max when I said I would be along later—if the pregnancy persisted. Max said he would phone me with any new developments.

For nearly three months now, Darwin had been implanting 16- and 32-cell embryos on a regular basis. But this first pregnancy was achieved with an embryo that was in transition between the 32- and 64-cell stage. Unfortunately, neither AnnaBelle nor Sparrow was in the proper phase of the menstrual cycle to receive the implant. Consequently, it was implanted in one of the other two surrogates, a young woman I will refer to as "M."

In little more than a week after the implant was performed, Darwin was aware of the possibility that M might be pregnant. An analysis of her urine indicated that a hormone associated with pregnancy was significantly elevated. Soon firmer pregnancy tests confirmed what Darwin suspected—and hoped for. M was definitely pregnant.

Darwin did not tell the girl this. The hormone injections by which he hoped to fortify the pregnancy were, he told her, simply a new phase in the study. When M missed her next period, she was told not to be alarmed, that this was a result of the injections, nothing more. He had decided it was best not to tell her any more until and unless the pregnancy gave signs of continuing.

It was just as well, because the pregnancy didn't last. The hormonal indicators of pregnancy as determined by urinalysis began to fade little more than three weeks after implantation. Soon all pregnancy tests were negative. Darwin was dejected again.

Max, though, when he told me this over the phone, sounded almost pleased. He had never cared for M. It wasn't that he actively disliked the girl; it was just that there was nothing about her to spark his interest. She had neither AnnaBelle's gushing energy nor Sparrow's mystery. Neither did the other surrogate.

After the failure with M, Max insisted that she be cut from the program. Perhaps, he argued, the fault was with the girl and not with Darwin's technique. The doctor could not argue too strenuously against that. And so M was eliminated. It was evident to me from many things Max said that he was increasingly of the mind that Sparrow must be the surrogate. She was assuming a place in his life almost as important as the clone itself.

Max had a conversation with Darwin about this during the October visit, and Darwin gave me some of the details later. He was convinced that Max was not merely infatuated with Sparrow, but was in love with her.

Not long after M was eliminated, so was AnnaBelle. I would have thought the other girl, in whom Max had shown so little interest, would be the next to go. But it was AnnaBelle. It seemed that she had, with her childish demands, finally annoyed Max beyond endurance. She was told by Mary simply that the study was coming to an end and that Max would no longer be present at the hospital. Max had Mary give the girl an expensive parting gift. And so from November on there were again only two surrogates, the minimum, as far as Darwin was concerned. He reluctantly agreed to settle for two, but only for a few months. If they still hadn't succeeded in five or six months, he warned, he would again require more surrogates—perhaps many more.

I suspected that it would be the other way around, that in six months Max's obsession with Sparrow would grow to the point at which she would be the *only* surrogate, with Max willing to wait years, if necessary, for the pregnancy to occur. Or perhaps, the thought came to me, he would even decide to marry the girl and father a child in the normal fashion if Darwin's efforts were not soon rewarded with success.

During the four-month period of November 1975 through February 1976 Darwin and Paul performed almost one hundred nuclear transfers, utilizing many egg donors. Of these, Darwin told me, almost all survived the fusion process for at least a short time. Better than 60 percent underwent some cleavage. Nine attained the 32-cell stage or better, and seven of these were actually implanted.

In mid-March, I received word from Max by telephone that Sparrow was "apparently" pregnant and then, a week later, that she was "definitely" pregnant. I flew to the facility at once.

Darwin was restrained, almost pessimistic. He could think of dozens of things that might go wrong. Those first several days of waiting beat any cliff-hanger of an election or murder mystery for suspense. I was all nerves. Darwin tried to deaden his with gloomy predictions. Only Max seemed truly elated.

Sparrow was not told that the early pregnancy tests were positive. But when she missed her period, she knew something was up. Was she pregnant? Darwin said she might be, but hadn't wanted to upset her. He was afraid that *her* nerves might abort a fragile "fix." Sparrow told him he should have informed her as soon as he knew of the positive tests. The news would not have upset her, she insisted, but instead would have made her stronger, steadier. -

Darwin thereafter presented a happy face to Sparrow but continued to warn the rest of us that our cliff-hanger could still end in "sudden death," as he grimly put it. The first several weeks, he said, were particularly critical. He was forever cataloguing the complications of pregnancy. Did we realize that about one in every six pregnancies was doomed to end in miscarriage or spontaneous abortion? At least half of all women who become pregnant, it was suspected, miscarry at least once in their lives. This could occur for any of a number of reasons—but usually it happened because the fetus was abnormal in some way and nature was eliminating it.

Later on there would still be miscarriage to think about—fetal death between the fourth and seventh months of pregnancy. Then there was the possibility of prematurity to plague us, should we get to the eighth month. Around 9 percent of the babies born in the United States, Darwin noted, were born prematurely, accounting for 70 percent of infant deaths during the first month of life. Thousands of those born prematurely didn't die but ended up deformed or mentally retarded. Half a million brain-damaged babies were born in the United States alone each year, he continued relentlessly. Prime contributors to prematurity and miscarriage were toxemias and placental insufficiency.

Darwin was an expert on fetal development and seemed almost to enjoy lecturing us on the hazards of pregnancy. He spoke of the placenta and its mysteries with awe. He called it a "physiological computer," one which he said put to shame anything IBM had ever concocted. This one-inch-thick hunk of tissue, quite unassuming in appearance, was "the quartermaster of pregnancy." It was through countless tiny, fingerlike projections on the placenta that the developing fetus obtained just the right balance of oxygen, food, and other life-support substances.

And it was a very sensitive organ, the placenta. Smoking, drinking, illness, even high altitudes or emotional upset could throw it off. Whenever a pregnancy was in trouble, whenever the growth rate of the baby slowed or stopped, placental insufficiency was usually the culprit.

Then there was toxemia, still something of a mystery disease, in which toxins would build up, sometimes abruptly, and enter both maternal and fetal bloodstreams. It was suspected that the placenta was the origin of these poisons. As it aged, the placenta sloughed off cells which in some cases apparently piled up. If the body couldn't dispose of these wastes quickly enough, the placenta could become clogged with them, reducing its efficiency and endangering both mother and child. These pernicious toxemias often didn't manifest themselves until late in pregnancy.

It was no wonder, Darwin added, that he and most of his colleagues continued to push the birth-control pill, despite a seemingly endless succession of reports on its adverse side effects. The fact was that the risks of pregnancy still continued to "vastly outweigh" the risks of pill-induced cancer, blood clotting, or other adversity.

In surveying the hazards of pregnancy, Darwin reminded us, he was talking about *normal* pregnancy—not one that began with a body-cell nucleus, a gutted egg cell, and a virally induced conception. "God knows . . ." he would say. And then, having scared the hell out of himself as well as us, he would talk hopefully of decompression and amniocentesis. Decompression to help discourage some of the darker possibilities and amniocentesis to discover them should they settle in anyway. Amniocentesis would not reveal everything, but at least it would point out any gross chromosomal abnormalities.

As for decompression, Darwin said he intended to begin using it immediately. He betrayed a mild suspicion, it seemed to me, about some of the South African data on decompression. Like so many of his peers he could be rigidly orthodox. Data that didn't issue from the United States, England, Germany, and perhaps one or two other countries, no matter how carefully presented, rational, and methodical it might be, never seemed quite good enough. I reminded him once that it was a South African who had achieved the first successful heart transplant. Ah, yes, he said, but then that surgeon was trained in Minnesota. Nevertheless, Darwin was by this time thoroughly sold on decompression—though mostly, it seemed, because of reports from *outside* South Africa, reports noting its beneficial effects on placental functions.[36]

Darwin had said that he would treat any pregnant surrogate as if she were a "habitual aborter," meaning given to spontaneous abortion or miscarriage, so Sparrow was moved from the room she had occupied in the hospital portion of the facility to a room off the laboratory. This had previously been used for storage. It was carpeted and repainted and made cozy. Sparrow didn't care much what it looked like as long as it had adequate light. And as long as she had her books. Max had some of the jungle chopped away from that side of the building so that more light could get through the window and into Sparrow's new room, and she was provided with two desks, one for reading and writing, the other for drawing and painting. She missed being able to go out for walks, but Darwin fretted around her like a mother hen, reminding her that she must be careful.

A decompression unit was moved to her room as well. Sparrow had no difficulty adapting to it; she mastered the controls in one session. It wasn't long until she was spending an hour a day in the thing. She said the sensations were not unpleasant; in fact she soon came to regard them as relaxing—and "good for meditation." As the pregnancy advanced, Darwin later told me, Sparrow would often say "The baby feels it" or "The baby likes it." She said the fetus became more active, kicking and thrashing about, during and after treatments.

I spent two weeks at the facility on this third visit, which turned out to be my last. Max, however, "commuted" every couple of weeks between the laboratory and Marin County and kept me constantly posted. By this time I had won the fellowship I had applied for and was engaged almost full time with that work, investigating the politics of cancer research around the world.

Even though I could not be at the facility during this period, I think I worried as much as the others about something going amiss in Sparrow's pregnancy. As we approached the fifteenth week of pregnancy—and all seemed still to be going well—our nervousness increased. A major test was about to be performed— one that could, we all knew, instantly dash any hope of success.

CHAPTER 30

There was really only one good technique available by which we might attempt to discover whether we had created a healthy baby —or a monster. This technique was known as amniocentesis, and involved analyzing fetal cells under the microscope. Darwin said that it should be performed at about the fifteenth week of pregnancy. At that point there would be ample fluid around the fetus. Some of this fluid would be sucked up into a syringe through a long needle and the cells floating in it analyzed. The idea was to look for chromosomal and biochemical abnormalities, any number of which, we knew, could mean disaster. It had been agreed from the beginning that, if there were any indication of deformity or mental defect, an abortion would be performed.

I think it safe to say that we all feared this test. I wondered how we would react if anything turned up abnormal. Nothing, not even Darwin's clinical horror stories, had held our excitement in check so powerfully as the prospect of this test. If we didn't get over this hurdle, it was "back to the drawing board." Prenatal diagnosis is still in its infancy, but amniocentesis, its principal implement, was all we had to go on—that and the traditional measurements of health or lack thereof in the womb, such as various hormone levels, the overall well-being of the mother, the rate of weight gain, fetal movements, and the like.

One thing we didn't have to worry about, Darwin noted, was whether the baby would turn out to be a boy or a girl, though amniocentesis could reveal that as well. For a short time I wondered—what if it turned out to be a girl? What if, by some circumstance, Sparrow was not recapitulating "virgin birth" but had become impregnated in the familiar way? It even fleetingly crossed my mind that Sparrow might have connived to become pregnant in order to succeed for the unseen couple, where Darwin and Mary had failed so many times before. I did not give voice to this errant fear.

What worried me far more was how Max and the others would react if amniocentesis signaled some equivocal danger—some only partial risk. Might there then not be a terrible temptation to proceed anyway—to take the risk and hope for the best? It would be difficult to give up what had been so arduously achieved. I wondered what my own reaction would be and decided that I *must* insist upon an abortion if there was any chance at all of producing a defective child.

I had discussed this with Darwin and Mary earlier. Darwin was of the view that there were some chromosomal abnormalities that would *not* be an indication for abortion. He noted that the much talked-about "supermale" abnormality would be one such example. He referred here to those individuals who are chromosomally abnormal (possessing an extra Y male sex chromosome) but mentally and physically all right. The idea that XYY males were excessively aggressive and prone to violence had apparently not been borne out by closer study. Neither Darwin nor Mary had seemed inclined to make any commitment until after the results were in. I pressed the issue no further but was prepared to do battle if necessary when the time came.

Darwin, I felt, had responded a bit unfairly when he said he hoped Sparrow, with her deformity, would not hear me talking about abortion. He added, too, that he had more to lose than I did if he were to deliver a monster. I guessed that Max and he had talked about this as well, and that Max had instructed him not to be hasty or "categorical," a word Darwin used several times in one of his conversations with me.

Amniocentesis and subsequent chromosomal analysis had been demonstrated effective for diagnosing not all but certainly a wide range of prenatal disorders. Performed with only local anesthetic, the procedure involved pushing a 4-inch needle with a very small diameter (about .5mm) through the abdominal wall and uterus into the amniotic sac. Thus situated, this needle could be used to suck up a bit more than a half ounce of amniotic fluid. There was some speculation in scientific circles that this removal of fluid might lower IQ by a few points, but if so, Darwin said, he expected decompression to more than make up for it. The fluid, once obtained, was placed in culture dishes and the cells it contained grown for later study.

In order to avoid pricking the fetus with the needle, Darwin said, he would first chart its location with the use of an ultrasonic scanning device. I was told that Sparrow watched with fascination as the sound waves that reverberated through her body were converted to visible light impulses on an oscilloscope screen and the dark image of her baby took shape before her eyes. The composite black-and-white picture of the scan revealed a recognizable fetal shape—a shadow figure enshrouded in the light that was reflected sound. Mary had explained the technique in detail to Sparrow, who had also been studying pictures of fetal development. When the girl saw the ultrasonogram of her own fetus, Max later told me, she referred to it as "the echo baby," a more apt description than she could know.

With the fetus and the placenta localized, Darwin was able to guide the needle into the amniotic sac without inflicting damage. Sparrow reported no discomfort. The cells suspended in the aspirated fluid were primarily sloughed from the skin of the fetus, Darwin said. The culturing process took time, and it was nearly two weeks before the cells were sufficiently developed for him to begin the chromosomal analysis. Cells were broken open and their chromosome pairs spread about on a slide. This cluster of chromosomes was then microphotographed. These *pictures* of chromosomes were next cut out and each individual pair aligned in a standardized fashion. The final arrangement was then itself photographed to produce the actual "karyotype" or chromosome pattern.

Abnormalities of chromosomal number could immediately be spotted, of course. Had there been an extra chromosome attached to the twenty-first chromosome pair, for example, we would have known at once that the baby, if permitted to be born, would suffer from Down's syndrome, better known as mongolism.

Numerous other abnormalities of chromosomal number were known to cause a variety of defined defects. The first thing Darwin looked for were these numerical anomalies. There were none.

The karyotype could also be used to detect structural abnormalities of chromosomes. Sometimes chromosomes broke and then mended with broken pieces attaching themselves at the wrong places. These structural defects were known as "unbalanced translocations" and they, too, could give rise to serious problems. Darwin next looked for such defects but found no missing or duplicated or misplaced chromosomal parts.

By culturing the amniotic cells for another two weeks, he was able to perform biochemical studies that enabled him to assess at least partially the metabolic health of the fetus. What he did here was check to see if the cells were producing various enzymes, the absence of which would indicate serious inborn errors of metabolism. Again, no abnormality was revealed. Nor was there any sign of two other fearsome possibilities, spina bifida and ancephaly, central nervous system defects that would either prove lethal or extremely debilitating. And the fetus *was* a boy.

I had earlier asked Darwin if he could use some of the fetal cells he obtained with amniocentesis to demonstrate that the baby was indeed Max's clonal offspring. But he said no, that other tests would have to be performed after the birth of the baby in order to prove that it was really the product of nuclear transfer. Max had indicated that he would conduct an independent investigation in that matter. But what he *could* do with some of those fetal cells, Darwin said, was to experiment with them in additional nuclear transfers, and if this pregnancy should, for any reason, miscarry, he would next attempt, not to clone Max directly, but to clone his fetal copy. The fetal cells, he felt, might prove far more responsive.

Max, keeping me posted by phone, told me that, with the results of the amniotic analyses in, even Darwin had difficulty acting pessimistic. Rather, as was so often the case with Darwin, he jumped for a time to the other extreme, taking the view that it had all been something of a breeze. He seemed quickly to forget that only a short time before he had been beating the drums of caution and pessimism louder than before—mainly in response to reports that the first pregnancy achieved by Steptoe and Edwards in their embryo-transfer work had suddenly ended in the ninth week. This had been the British researchers' first real hit after some forty implantation attempts. We knew how disappointed they must have felt when their pregnancy began to dissolve before their eyes.

Darwin was pleased by only one aspect of the Steptoe-Edwards experience. That they had come as near success as they had could be accounted for by the fact that they had finally begun to make attempts at implanting advanced rather than merely 8- or 16-cell embryos.[37] Darwin took an I-could-have-told-them-so stance.

Actually, the cause of the miscarriage in Britain was unlikely to apply in our case. The Steptoe embryo had implanted in the oviduct of its recipient rather than in her womb, and the scarred and otherwise damaged nature of the tube appeared responsible for "holding" it there, where it could not grow normally. The British researchers, in the wake of this failure, decided that some of their tube-damaged candidates for embryo transplantation might first have to undergo surgical removal of their tubes. It was a real setback for them—"political" as much as scientific, for the opponents of their research were now quick to point out how easy it might be to produce a deformed or retarded child, in short, a laboratory monster.

Counterbalancing the bad news from Britain to some extent had been encouraging news from Texas, where a healthy male baboon, researchers reported, had been born as a result of an embryo-transplant effort. They had removed an embryo at the morula stage from the uterus of the donor female and then transplanted it into a surrogate baboon. The healthy baby was delivered in due course by cesarean section, used as a precaution against delivery damage. What was particularly interesting to us

197

was that the Texans had succeeded after only eleven implant attempts.[38]

For a time at least, though, the news from the rest of the world had little impact upon us. The good results from the chromosome tests buoyed us so much that we began talking as if human cloning might already be regarded as routine. We all began finding compelling new uses for it.

Mary knew a couple who, with her help and great difficulty, had overcome an infertility problem of long standing. They had finally conceived a child only to nearly lose it to prematurity. There had been a long and costly battle to keep the baby girl alive. They succeeded in this, and the child, while she didn't exactly thrive, had reached the age of four when she was badly injured in an automobile accident. She had thereafter lapsed into a coma and seemed now to be steadily declining toward death. There was almost no chance that the woman could become pregnant again, Mary said. Both the woman and her husband were "nearly suicidal" at the prospect of losing their daughter. If ever there was a valid place for human cloning, Mary said, this was it.

Then Darwin, Max told me, spoke of one of his former professors—now an old man—who had, after his first wife died, married one of his students, a girl young enough to be his granddaughter. The professor would die within a few years, Darwin was certain. He thought it would be wonderful if the girl could know her husband as he had been "as a child and a young man." In other words, she could perhaps carry and then marry the reproduction of her late husband. Or perhaps she'd be satisfied with simply "mothering" the man who had, in a sense, "fathered" her. It was a bit exotic for my taste, but I credited Darwin with good intentions. At least none of us suggested cloning King Tut's mummy.

Max, however, began hinting that he might eventually be in the market for another copy of himself, and that, I think, was the factor that brought us all rushing back to earth. Thereafter we were more concerned with establishing limits than in extending the frontier. We would all later agree, at least Darwin, Mary, and I would, that no further attempts would be made until and unless, at the very earliest, everything appeared to work out with the present effort. For one thing, we still had several months to go before Sparrow would deliver. Even then, something undesirable might emerge. Amniocentesis, we all knew, was powerless to uncover a multiplicity of possible defects. And even if, by all the standard measures, the baby was physically and mentally normal, we still could not predict what unique tribulations his special station in life might visit upon him.

It was far too early even to think about a repeat performance. There were some earnest deliberations over the telephone aimed at defining the conditions under which another attempt might be made, but we were soon hopelessly mired down, and we decided that we would play it cautiously, by ear. Meantime, nothing would happen until after the baby was born—probably not until well afterward.

Max didn't seem to mind that we were less than enthusiastic about churning out additional copies of him. His own mind was far from made up on this anyway; he was simply blue-skying with the rest of us. His main interest at this point was Sparrow. He spent more and more time in her company.

By the end of the summer, I was told, she had been informed that there was no "couple." Sparrow was too smart, Max felt, to go on buying this story—or any modification in which Max himself didn't figure prominently. Might she accept a version that the wife had died and that Max was the bereaved husband? Max thought not—or perhaps he simply realized he couldn't play the part convincingly.

Yet he was still not willing to tell her the whole truth. That would be premature, he felt, perhaps by several years. Instead he told her that he was a very wealthy man who had no heir and that the child she was bearing was destined to become that heir. More than that, he told her, he could not say; she would have to trust him. Sparrow, he told me, looked at him closely for a while and then said that in that case *he* would "very much" have to trust *her*. Beyond that, she betrayed no surprise at this revelation.

CHAPTER 31

Sparrow left the hospital shortly after the amniocentesis was completed. She was moved to Max's house some miles from the facility. Mary, who already lived there, was in full-time attendance with Sparrow, and Darwin paid frequent visits. One of the decompression units was taken to the house as well. The move was aimed at curtailing any talk or gossip about Sparrow's pregnancy at the hospital. So far no one outside of our own circle knew what had happened. Besides, it was more comfortable at Max's house. Sparrow did not relish the idea of being cooped up and hidden from view at the hospital for the remainder of her pregnancy, and the grounds around the house afforded ample privacy and room to roam.

About this time Sparrow decided it would be wise to call upon her aunt, who would be worried now that her niece was no longer at the hospital. The aunt had been very happy for Sparrow in her situation at the hospital—if also somewhat mystified. Now, Sparrow said, she must explain to the aunt that she was going to another facility for further study and would not be seen for some time. She insisted that the meeting take place at the aunt's hut, where the older woman would be more comfortable.

Darwin was opposed to the visit, warning that "anything could happen." Sparrow might eat something or drink contaminated water, he fussed; she might fall; the aunt might persuade her to stay on with her and resume her old life. Sparrow laughed at his fears, and Max said it was all right. So Sparrow paid the visit with a rather comical "security force" in attendance.

There was a car hidden down the street from the aunt's hut, and Roberto, dressed "down" in an effort to blend in, skulked about outside. It all went off successfully.

Darwin and Paul found ample work to occupy them at the hospital during this waiting period. It was not long until they were engaged in some research of their own design—funded by Max. The project was not over until there was a live and otherwise satisfactory birth. If the worst of our fears were realized, if the baby emerged badly defective, it was possible that they might start all over again, though I at least would be adamantly opposed to a repeat effort. I suspected that Darwin would not take much persuading on this score, either. About Max I was not so sure, but anyway, this was something that could be dealt with only when and if it happened; in the meantime I hoped for the best.

I did worry quite a lot, though, about what would happen after Mary and Darwin left the area, after the baby was born. The quality of health care in the local area had been considerably enhanced by those two—and by the infusion of funds into the hospital, an infusion occasioned by their presence. It would seem cruel and unfair if all that should suddenly evaporate, leaving heightened expectations unfulfilled. We had some discussions about this, and Max agreed to bring in replacement personnel when appropriate.

Sparrow, meantime, had little difficulty filling her time. With a new language in her grasp and all the books she could read, a whole new history, a whole new culture was unfolding for her. Also, Max taught her to play chess—and suffered his first defeat at her hands only two months later. He would often sneak glances, he told me, at a painting of him she was working on. She continued to work on it for some months, and her changing conception of him, he felt, was reflected in the painting, which seemed to be undergoing constant metamorphoses. He claimed that, after she found out that the child she was carrying would be his heir, her portrayal of him became both "more handsome" and "a bit more sinister."

Max was secretive about his plans for Sparrow. He talked of marrying her; he also talked of adopting her. At this writing I am not sure whether he ever did either. He said that this need not concern me, that in the event of his death, provided the offspring was still legally a minor, I and others would be apprised of all pertinent details. In any case, there was no doubt that whether he lived or died, both Sparrow and the child would ultimately know the full truth.

From what he said to me on one occasion, I was left with the impression that he foresaw and perhaps even desired a lifelong relationship between Sparrow and his copy, one that might more closely approximate that of man and wife than of mother and child. It was thus important to him that they be assured they were in no way related. The difference in years between them would be substantially less than the difference between Sparrow and Max. And if Sparrow did fall in love with Max, which seemed certainly to be a possibility if not already a fact, it did not seem unreasonable to guess that some day she might similarly love his copy, once Max was gone and the boy had grown up.

It was decided well in advance of delivery that Mary would be retained by Max after the birth of the child. She was devoted to Sparrow—and to Max—though I wondered how the two women would ultimately divide their duties. Perhaps, I thought, Mary would end up the primary maternal influence and Sparrow would evolve into something more remote and thus, at the same time, more accessible as a future romantic interest. But Sparrow was very strong, with a mind of her own. Of one thing I was certain: it would be intriguing, whatever happened.

The pregnancy advanced uneventfully. For that we were all most grateful. We had all assumed, and I think Max himself had always intended, that the baby would be delivered at the hospital, where it had been conceived. Then, "at the last minute," as Darwin put it, Max decided that the child must be born in the United States. Although he did not tell me exactly why, I could see many advantages in Max's heir's being born an American citizen. Max had obviously found some legal way to bring Sparrow to the United States.

Darwin was troubled by this new plan. Actually, it was not the last minute. Sparrow was not expected to deliver for two and a half weeks, but Darwin said he would not be responsible for an eleventh-hour miscarriage. Again, he warned, anything could happen. What if Sparrow went into labor prematurely—on the plane? It was unthinkable he said. But Max was determined, and Sparrow, once apprised of the plan, equally so. After all, Max said, it was perfectly possible to have a baby on a plane. I think he halfway liked the idea. Anyway, it was not as though the plane would be a commercial jetliner; it would be one of Max's own cargo jets.

I tried to imagine them all huddling aboard it in the dead of night, tiny Sparrow with her swollen stomach, Max hovering over her, Mary and Darwin fretting along behind. The plane flew first to Mexico and then to California. It was December, and Sparrow had hoped to "feel snow" for the first time in her life. Max was tempted to go by a route that would make this possible, but Darwin objected, and this time so did Mary; the sudden climatic change might be dangerous.

Max had made arrangements whereby Darwin could deliver the baby in a small hospital. The decompression unit made the trip, too, and the plan was to deliver without anesthesia. Both Max and Sparrow had read some books on the subject and were converts to natural childbirth; Max said he did not want his heir coming into the world "drugged."

Sparrow was moved into a motel room, very plain but in the vicinity of the hospital, and I was at last able to join the group. We made jokes about being holed up, in hiding, and so forth. It was a tense period, particularly for Darwin, who did not travel so well and found California in December—at least as viewed from a rather inelegant motel—not at all to his liking.

Sparrow's contractions started a couple of days earlier than expected. Max called me from the hospital in the early morning

203

hours. The others were already there, he said; the contractions had come on so rapidly and so strongly that they had all rushed off, leaving me behind. I dressed and drove my rented car to the hospital, where Sparrow was already in decompression, looking quite content.

In fact, Darwin said, she had been in labor for some time, experiencing contractions that would have panicked most women. But Sparrow had been reluctant to awaken Mary, who shared her room, since she did not think the pain was yet enough to signal delivery. We had complete privacy, and all of us sat with Sparrow as she reacted to each contraction by escalating the negative pressure on the decompression unit to minus 100 millimeters of mercury or more. Only twice did she show any visible sign of discomfort, and that, she said, was her own fault, because she wasn't decompressing quickly enough or long enough.

We were all curious to know how long Sparrow's labor would last, particularly since she had used decompression from early pregnancy on. The South Africans had observed that the earlier decompression was used the speedier and less painful labor was likely to be, although they found that pain relief could be significant even in those women who were first introduced to decompression during labor itself, provided high enough negative pressures were used.

In addition to lifting the abdominal muscles up off the uterus, it was believed that decompression also worked to relieve pain and shorten labor by encouraging the Braxton Hicks (or so-called painless) contractions that tone up the uterus and soften the cervix early in pregnancy. Thus decompression could be thought of as "training" the uterus for the task of labor, bringing it up to optimal performance levels. A group of Canadian researchers had found that 50 percent of their first-time decompression mothers had first-stage labors of under three hours; far more than half of these women got "good" or "excellent" relief from pain during labor and required no anesthesia.[39] None of the women in this survey had used decompression nearly as long as Sparrow had.

Two hours and forty minutes after she had began decompressing, by Mary's calculation, Sparrow was ready to deliver. The second stage of labor—which neither Max nor I witnessed, having been forbidden by Sparrow to do so—was over in a matter of minutes. The baby presented head first, without difficulty, appeared immediately normal, cried robustly and was given a near-perfect Apgar score by Darwin. (The Apgar Scoring System, developed by Dr. Virginia Apgar of the National Foundation—March of Dimes, is now almost universally used to evaluate a newborn infant's health, taking into account such things as muscle tone, color, and reflexes.)

Max had wanted to film the birth, but Sparrow objected. It would be immodest, she said. But she did permit Max to place a tape recorder in the delivery room to catch the baby's first cry. Max was permitted into the delivery room a short time after the baby was born, and I was allowed in about twenty minutes after that.

Sparrow said that she wished the baby had come at Christmas—still two weeks away. Max was delighted that it had happened in 1976—his contribution to the Bicentennial, he said. Darwin was beaming. Mary looked almost beatific. Max was sitting on the edge of Sparrow's bed. She was holding the baby in a small blanket to her breast.

It was not, I thought, exactly the nuclear family. But it was a thrilling sight, this old man, this young girl, this strange baby. I wondered what this wrinkled little creature could see. I wondered what he might know. I wondered if he would be brave.

AFTERWORD

One year after the birth described in this book, Max's primary concern continues to be for his privacy and for the privacy of his new "family." Max, Sparrow, and the child are together—and well. Blood analyses and other tests involving various histocompatibility factors have demonstrated to Max's complete satisfaction that the child is indeed his clonal offspring. I have some reason to believe that Sparrow has now been apprised of the origin of the baby to which she gave birth.

Max is aware that I am writing this book and continues to be opposed to it, although I think now with lessened force. If he finds, as I am confident he will, that it does not reveal his identity, I believe that ultimately he will not be displeased by its publication. Darwin, too, may find it useful in a sense, for I know he had been torn between wanting both to protect himself and to honor his commitment to Max and wanting to lay claim to his accomplishment. Public reaction to the book will be of great interest to both men and may well influence some of their future actions. I am confident that, in the fullness of time, complete documentation of this accomplishment will be forthcoming. Many more details may emerge within the next few years.

In the meantime, I entertain absolutely no expectation that

anyone, scientist or layman, will accept this book as *proof* of the events described herein. I am fully cognizant of and fully respectful of the methods by which scientific data must be conveyed. I hope, however, that many readers will be persuaded of the possibility, even the probability, of what I have described and will benefit by this preview of an astonishing development whose time, at least in terms of some of the emotional and ethical issues it raises, has apparently not quite yet come. And if my book, for whatever reason, increases public interest and participation in decisions related to genetic engineering, then I will be more than rewarded for my efforts.

There will be those who honestly doubt the veracity of the book, others who fervently believe it. With those individuals, I can have no quarrel. But I am also aware as I conclude this book that there will be those who falsely doubt; that is, those who falsely claim disbelief. Some will do so out of the same sort of professional jealousy that has been visited upon the Shettleses, Bevises, Barnards, and so on; others, of higher principle, will express doubt because they will sincerely fear that, if it is accepted that a man has been cloned, then the public, in panic or disgust, will turn against the sciences or regard themselves as somehow degraded or depersonalized.[40] With these I must take issue, as I did in the beginning of this book.

And still others, I fear, will falsely believe in order to cry wolf and thus induce the panic or disgust by which they hope to halt the experimentation they are categorically opposed to, for reasons I cannot consider sound. These too I will continue to take issue with.

In summary, I expect the response to this book to be complex, perhaps harrowing, but finally, I hope, illuminating.

David M. Rorvik

San Francisco
January 1978

NOTES

PART I: MORALS

1. For a summary of some of these views, see Edward O. Wilson, *Sociobiology* (Harvard University Press, 1975).

2. James Watson, "Moving Toward Clonal Man: Is This What We Want?" *Atlantic Monthly,* May 1971.

3. "Toward Preselected Sex," *Science News,* August 3, 1968.

4. Willard Gaylin, "Frankenstein Myth Becomes Reality," *New York Times Magazine,* March 5, 1972.

5. "Embryo Transplanting May Alter Cattle Industry," *Kansas City Star,* August 1, 1971; "Sci-fi on the Range," *Newsweek,* September 4, 1972.

6. Bentley Glass, "Genetics and Surgery," *Bulletin of the American College of Surgeons,* November 1972.

7. Joshua Lederberg, "Experimental Genetics and Human Evolution," *Bulletin of the Atomic Scientists,* October 1966.

8. "Doomsday: Tinkering with Life," *Time,* April 18, 1977.

9. David M. Rorvik, "The Embryo Sweepstakes," *New York Times Magazine,* September 15, 1974.

10. *Protection of Human Subjects: Policies and Procedures,* National Institutes of Health *Federal Register,* vol. 38, no. 221, Part II, November 16, 1973.

11. Walter Sullivan, "Frozen Mouse Embryos Grow in Foster Mother," *New York Times,* August 15, 1972.

12. L. E. Kass, "New Beginnings in Life," in Michael P. Hamilton, ed., *The New Genetics and the Future of Man* (Eerdmans, 1972).

13. J. Fletcher, *The Ethics of Genetic Control* (Doubleday, 1974).

14. D. Callahan, "New Beginnings in Life: A Philosopher's Response," in Hamilton, *The New Genetics and the Future of Man.*

15. "Reports Human Egg Fertilized in Test Tube," *New York Post*, January 13, 1961. See also Albert Rosenfeld, *The Second Genesis* (Vintage Press, 1969), and Ted Howard and Jeremy Rifkin, *Who Shall Play God?* (Delacorte Press, 1977). Petrucci's claims were widely discounted by scientists around the world, even though the Italian's credentials were all in order. Soviet researchers, however, invited Petrucci to Russia, where for two months he apparently shared his knowledge with scientists at the Soviet Institute of Experimental Biology. The Soviets were sufficiently pleased with what Petrucci told them that they awarded him a medal for his scientific work. Five years later the Soviets claimed to have grown more than 200 human embryos in vitro—some for longer than two months. Perhaps the unpopular nature of this work led to a blackout of further news of such developments in the Soviet Union.

Meanwhile, among thoroughly documented reports of mammalian gestation outside the womb, that of Dr. Yu-Chih Hsu of the Johns Hopkins University School of Hygiene and Public Health continued to define the state of the art. Dr. Hsu succeeded in growing mice embryos to the heart-beating stage in a laboratory container. ("Mouse Embryos Grow in Absence of Uterine Tissue," *Ob.Gyn. News*, May 15, 1971.) Still, if the 1966 report of Dr. Pyotr Anakhin of the Soviet Academy of Medical Sciences in Moscow was true (asserting that the Russians had surpassed Petrucci), then the Soviets could certainly have attained by now the ability to clone a human being.

16. Vincent and Margaret Gaddis, *The Curious World of Twins* (Hawthorn Books, 1972).

17. Ian C. Wilson and John C. Reece, in *Archives of General Psychiatry*, October 1964.

18. Gunther Stent, "Molecular Biology and Metaphysics," *Nature*, April 26, 1974. Biologist Stent had it that all the great visionaries had populated their ideal societies, the ones that seemed perpetually to be on the drawing boards, with angelic characters, men and women of only the highest attainments. Heaven was supposed to be like that, too, but now that it was possible—or would be before long—to create the "perfect" society by abandoning the "old-fashioned genetic roulette of sexual reproduction [in favor of] populating the Earth with identical replicas of carefully chosen, ideal human genotypes," the visionaries were running like hell, horror-struck by the prospect of it all. Stent seemed to be arguing, though with tongue intermittently in cheek, for the conversion of the West to a "purely pagan metaphysics" that he said would obviate some of the inconsistencies, contradictions, and hypocrisies of the Christian ethic, at least as practiced by said horror-struck visionaries.

19. Horace Judson, "Fearful of Science," *Harper's*, March 1975. In this piece Judson quoted theologian Paul Ramsey at length, obviously identifying with his categorically negative viewpoint on embryo transfer and cloning. Dr. Edwards was faulted for what Judson called his "bumptious confidence." An unnamed source was quoted as saying "Edwards only wants to be a Christiaan Barnard." Edwards, Bevis, and the others, Judson continued, "behind their unctuous language display a stubborn incomprehension of what is being said to them." Judson recounted an exchange that had taken place at a Ciba Foundation symposium on embryo work in London. Dr. Bevis had said, "How can any society that accepts termination of pregnancy quibble about giving life to a fetus? We are not creating life. If we can kill the fetus—and this seems to be expected and accepted

—why can we not 'put it together'?" In response to this "unctuous language," Judson quoted British Nobelist Dr. Max Perutz, who had replied to Bevis, "If you kill a fetus nothing happens, there will be no child, but if you 'put together' a child, you might produce an unhappy individual. The responsibility is of a different kind." Darwin wrote in red ink in the margin of this article, "Nothing happens, my ass! You merely kill an individual. How does the potential for unhappiness differ whether putting together a fetus in the laboratory or in bed?" He had underlined the words "unctuous" and "stubborn incomprehension" and had connected them by arrows to the names Judson and Perutz.

20. Editorial, *Science*, October 25, 1974.

21. Lewis Thomas, "Notes of a Biology Watcher: On Cloning a Human Being," *New England Journal of Medicine*, December 12, 1974.

PART II: METHODS

1. R. C. W. Ettinger, *Man into Superman* (St. Martin's Press, 1972).

2. H. J. Muller, "Human Evolution by Voluntary Choice of Germ Plasm," *Science* 134 (1961):643–649. See also Muller's *Studies in Genetics: The Selected Papers of H. J. Muller* (Indiana University Press, 1962).

3. E. S. E. Hafez, personal communication. See also David M. Rorvik, *Brave New Baby: Promise and Peril of the Biological Revolution* (Doubleday, 1971).

4. D. M. Rorvik, "Babies for Sale," *Good Housekeeping*, September 1976.

5. Roger W. McIntire, in *BioScience*, January 15, 1971.

6. Joshua Lederberg, "Experimental Genetics and Human Evolution," *Bulletin of the Atomic Scientists*, October 1966. For more recent views see Lederberg's "DNA Splicing: Will Fear Rob Us of Its Benefits?" *Prism*, November 1975.

7. Robert Bahr, "A New Ethical Question: Head Transplants?" *Science Digest*, May 1977.

8. Quoted in D. M. Rorvik, *As Man Becomes Machine: Evolution by the Cyborg* (Doubleday, 1971).

9. D. M. Rorvik, "Behavior Control: Big Brother Comes," *Intellectual Digest*, January 1974, and "Bringing the War Home," *Playboy*, September 1974.

10. Ernest Borek, *The Sculpture of Life* (Columbia University Press, 1973).

11. J. Rostand, *Can Man Be Modified?* (Basic Books, 1959).

12. D. M. Rorvik and O. S. Heyns, M.D., *Decompression Babies* (Dodd, Mead, 1973).

13. The discovery came about in this fashion. In desperation, Heyns and co-workers finally decided to use a curare-related drug to paralyze temporarily the abdominal muscles of a pregnant patient who seemed to be in doubtful labor, with her contractions so weak and poorly established that they were not likely to lead to delivery soon. The doctors were thus able to put the patient's muscles out of commission and get some good uterine readings, but what stunned them was the rapidity with which, once curarized, she delivered. This woman was a primipara—a woman having her first baby—for whom labor would typically last about fourteen hours. This woman delivered in six. Several subsequent patients given curare confirmed the effect—many delivering their first babies in two or three hours. The curare was doing what many natural-childbirth techniques seemed to do:

211

it relaxed the abdominal muscles and permitted the uterine contractions to work to the fullest. But curare, of course, was hardly ideal, and many women could not sufficiently overcome their fears to take advantage of natural childbirth. Heyns wanted a safe, reliable means of removing the abdominal interference in *every* case. There had to be some means of lifting those big muscles up off the uterus. It wouldn't be easy to find, he knew, for apart from the fear-induced tension those muscles could bring to bear, there was also pressure from the atmosphere (14.7 pounds per square inch at sea level) putting a load of at least *half a ton* on the abdomen and uterus of the typical woman.

14. See, for example, J. A. Blecher and O. S. Heyns, "Abdominal Decompression in the Treatment of the Toxemias of Pregnancy," *Lancet* 2 (1967):621, and A. Dolezal, "Increase in Uterine Blood Flow in the First Stage of Labor During Abdominal Decompression as Measured with ^{131}I-HSA," *American Journal of Obstetrics and Gynecology*, January 1970. Despite these publications and even after an American physician won first place with an abdominal-decompression display in the Scientific Exhibits of the American College of Obstetricians and Gynecologists in 1971, there still was little interest in decompression evidenced among American doctors. I attribute this to professional inertia; doctors—and especially specialists—became set in their ways and protective of their professional prerogatives.

Decompression is something like natural childbirth in that it gives back to the patient some of the responsibility in pregnancy and childbirth. The patient can operate the decompression suit herself (and, indeed, in South Africa, many women take the suits home with them) and remain alert and conscious through the delivery of her child. Doctors like to be in total control; in the mind of the baby doctor, the best patient, alas, is still all too often the unconscious patient. If women are ever to benefit from decompression on a large scale, they would probably have to set up their own decompression study groups, classes, and clinics and bring decompression kicking and screaming into the real world the same uphill way they had championed natural childbirth.

15. See, for example, R. G. McKinnell et al, "Transplantation of Pluripotential Nuclei from Triploid Frog Tumors," *Science* 165 (1969):394–396. Note, however, that some authorities have questioned whether it was really cancer-cell nuclei that were transplanted; see J. B. Gurdon, *The Control of Gene Expression in Animal Development* (Harvard University Press, 1974), in which Gurdon states, "There is no proof that the cells from which these nuclei were transplanted were tumour cells and not non-malignant host cells present in the tumour tissue. To be sure that a cancer cell nucleus is transplanted, it would be necessary to clone a single cell, demonstrate that some of the cloned cells are malignant by injecting them into a susceptible animal, and then transplant nuclei from the rest of the clonal cells."

16. See, for example, B. Mintz and K. Illmensee, "Normal Genetically Mosaic Mice Produced from Malignant Teratocarcinoma Cells," *Proceedings of the National Academy of Sciences* 72 (1975):3585–3589.

17. Stanley Cohen, "The Manipulation of Genes," *Scientific American*, July 1975. Dr. Cohen is one of the leaders in the new DNA work. It was after reading this article that Max began populating his conversations with such terms as "plasmids," "restriction enzymes," "ligases," and the like. The Cohen article seemed to bring many of the new breakthroughs in genetics into focus for Max. It had been possible—though not easy—

to combine the genetic material of different species since the late sixties, but it wasn't until scientists learned some tricks from lowly bacteria that this task became considerably easier —more controllable and more predictable. Bacteria are single-cell creatures that reproduce by simple division. Without the benefits of crossbreeding, these lowly animals would seem to be ill suited for evolution or even for minor adaptive change, and indeed most bacteria have remained unchanged throughout their known histories. Yet some of them have persisted even in the face of powerful human efforts to eradicate them. Some, for example, have acquired the ability to break down penicillin molecules so that they are harmless to the bacteria. This could happen only if some genetic exchange had taken place, conferring this remarkable ability upon these bacteria.

Scientists discovered that the vehicle for this exchange is the "plasmid," a tiny loop of extra-chromosomal DNA that can be passed from one bacterium to another when they brush up against each other. Science was not long in envisioning some very interesting uses for these bits of "genetic loose change," as some called them. Maybe, scientists thought, we could piggyback some messages of our own onto these plasmids—add a meaningful postscript here or there or perhaps edit the bacterial instructions to suit our own ends. Soon a variety of "restriction enzymes" that possess the ability to cause DNA chains to break at specific points were discovered, again by observing bacteria in action. Plasmids were found to contain the instructions that induce cells to create these enzymes. They were useful to bacteria as a means of disrupting the DNA of invading viruses. In conjunction with "ligases," previously discovered enzymes which can glue together— rather than sever—loose ends of DNA, plasmids and their restriction enzymes could be used to literally "cut-and-paste" DNA sequences to order. This molecular editing became miraculously simpler with the discovery by Robert N. Yoshimori, in the laboratory of Herbert Boyer at the University of California, San Francisco, of a restriction enzyme which in one stroke could cut out specific sections of DNA and leave behind two sticky ends upon which to insert a new segment of DNA. Then Cohen and colleagues discovered a plasmid that was uniquely suited for introducing new messages right down to specific genes. It was now possible to combine genes of unlike species, if one desired, without interfering with their replicative ability. Once introduced into the *E. coli* bacterium, for example, the plasmid-borne message would be incorporated into bacterial DNA and reproduced millions of times over in a few hours. This was the process of molecular cloning.

18. All sorts of substances, scientists promised, could be produced through recombinant DNA work—insulin for diabetics, growth hormones for the stunted, antibodies to fight numerous diseases, enzymes for the treatment of those deficient in them, and so on. Instead of tediously extracting such substances from ground-up tissue, for example, scientists hoped now to be able to insert human genes into bacteria and thus get the bacteria to do all the work, producing the needed substances under the direction of the newly added human DNA. It was not long before some of these hopes began to show signs of being fulfilled, as researchers moved closer to the molecular production of insulin and a brain hormone that governs many physical functions. In the case of this hormone it was possible to produce, in a very short time, as much of the substance as had previously been obtained only by grinding up some *half million* animal brains.

19. By the spring of 1977 the controversy swirling around recombinant DNA work

had reached such a pitch that students and others opposed to the work stormed the stage of a forum of the National Academy of Sciences in Washington, D.C., where proponents of the research were discussing their plans. The insurgents unfurled a huge banner that read, "We Will Create the Perfect Race—Adolph Hitler, 1933," and chanted not "We shall overcome" but "We shall not be cloned." In some quarters, it was evident, recombinant DNA was feared and hated. Opponents of the work warned that we might create deadly new life forms. An experiment had already been contemplated in which the *E. coli* bacterium, a normally benign inhabitant of the human gut, would receive some of the genes of the SV40 monkey virus in an experiment designed to shed light on this virus's apparent ability to encourage normal cells to become malignant. It might thus be possible to discover *which* gene, if they were transplanted one by one into *E. coli*, caused cancer. A worthy goal. But there were problems. What if the altered *E. coli* were to proliferate with its new SV40 genes intact, rapidly spreading through the population, seeding a biological catastrophe that might become evident only years later in the form of an epidemic of literally malignant dimensions? It didn't seem likely that the edited *E. coli* could confer cancer upon man, but the mere suggestion that it might was so frightening that the man who came up with the idea, Paul Berg, a Stanford biochemist, decided to cancel his experiment.

That was in 1971. Then came the breakthroughs in Boyer's and Cohen's labs (see note 17), and experiments like the one Berg had contemplated became temptingly easy. Other scientists became concerned, and soon 140 of them met at the Asilomar Conference Center in Pacific Grove, California. The recommendations of this group were later incorporated into the National Institutes of Health *Guidelines for Research Involving Recombinant DNA Molecules* (June 23, 1976). These recommendations were aimed at reducing the possibility of someone's creating a new virus or a bacterium capable of producing virulent new toxins or of inflicting cancer or other diseases upon the population. They called for various "containment" laboratories in which the more dangerous work would be carried out. The Asilomar conference was widely hailed as the most splendid display of scientific responsibility since 1939, when a group of atomic scientists agreed not to publish further details of their work lest they benefit the Germans. In fact, however, as men like Berg have conceded, most of the participants at Asilomar were self-serving and wanted only a free hand to carry out their experiments. It was only after a group of attorneys had been called in to discuss the legal ramifications of, for example, a virulent, rampaging new life form loosed upon the world that any sort of cautionary consensus was achieved. The lawyers had put it in dollars and cents. Multimillion-dollar lawsuits would almost certainly result if anybody "messed up," they warned.

Molecular biologist Jonathan King of M.I.T. has said that the Asilomar exercise in "self-policing" was on a par with "having the chairman of General Motors write specifications for safety belts." Nobel Prize winner George Wald has called for a complete halt to all recombinant work. "Our ignorance is profound," he has stated. And noted Columbia University biochemist Erwin Chargaff asks, "Have we the right to counteract, irreversibly, the evolutionary wisdom of millions of years in order to satisfy the ambition and the curiosity of a few scientists?" Cohen has responded by saying that it is Chargaff's "evolutionary wisdom that gave us the gene combinations for bubonic plague, smallpox, yellow fever, typhoid, polio, and cancer." (For more on the dangers of recombinant DNA, see note 40.)

20. The oil-eating bug is more than mere fancy. General Electric has already devised a method whereby a plasmid (see note 17) could introduce into the *Pseudomonas* bacterium the instructions that would enable it to digest most of the hydrocarbons in crude oil. It is also considered possible to induce algae that use only sunlight for energy to release vast amounts of hydrogen, which in turn could be used as pollution-free energy by man.

21. *Scientific American*, March 14, 1977.

22. As in note 40, the corporation in question is General Electric. For more on this plan see note 40. Max's entrepreneurial enthusiasm over recombinant work has been echoed by another businessman, Dr. Ronald Cape, president of the newly formed Cetus Corporation. Dr. Cape predicted that in the next several years "biology will replace chemistry in importance in this country." The power of genetic engineering, a Cetus memo asserted, "cannot be exaggerated . . . a new industry with untold potential is about to appear." Among Cetus advisers: Dr. Joshua Lederberg, Dr. Stanley Cohen, and other leaders in the field. Upjohn, Miles Laboratories, Hoffman-LaRoche, Eli Lilly—all of the major pharmaceuticals were getting into recombinants, Max told me. It wouldn't be long, he felt, until General Motors, General Foods, and others would follow "in a big way."

Max was also excited about the commercial possibilities of cloning entire organisms (as opposed to molecular cloning). Early in 1975 he told me about a plan to "clone a whole forest from a handful of cells." Because of his interests in agriculture and silviculture, this idea filled him with excitement. The plan was to start out by trying to clone the Douglas fir, a tree highly valued for its sturdy lumber. Trees, Max went on, were primitive things; they hadn't been improved the way so many grains and plants had been, through hybridization and other manipulations by man. Trees were difficult to deal with because they were so slow to grow. You could try to get an exceptional tree through selective breeding and the like, but you'd have to "wait your lifetime," as he put it, before you could tell if it really was exceptional and also before you could get enough seeds to start even a small forest. But with cloning, he said, you could take the tree of your choice, isolate hundreds, thousands, millions of its body cells, grow them in nutrient solutions like those used by Dr. Steward in the cloning of carrots, and then plant the resulting shoots to produce a forest of consistently superior and largely identical trees, depending for variation only upon inconsistencies in the soil and atmosphere. Trees could be selected for cloning on the basis of pest resistance, speed of growth, size, strength, and so forth. This plan, which was under study at the Oregon Graduate Center, a research institute, Max said could revolutionize forestry. (For a report on some of this work, see Tsai-Ying Cheng and Thanh H. Voqui, "Regeneration of Douglas Fir Plantlets Through Tissue Culture," in *Science*, May 16, 1977.)

Max felt that agriculture could go the same route; carrots, tobacco, parsley, barley, soybeans, endive, asparagus, and other plants had already been cloned. So had a number of valuable flowers, such as orchids. Cloning, it seemed, provided the perfect method of making multiple copies of unique orchids obtained sometimes by luck, other times by the most painstaking manipulation of the pollinating variables. With cloning, even a chance beauty could now be easily reproduced—not approximately but precisely, Max said, reaping the breeder a tidy sum of money. This orchid-cloning process was patented, and some specimens selling for up to $100 apiece carried the warning to other breeders: "Asexual Reproduction Is Forbidden." Max indicated that he was looking into the patent situation on other plants and flowers and was beginning to think about doing the same with various animal stocks. What seemed farfetched now, he said, would look "foresight-ful" before long.

23. *Guidelines for Research Involving Recombinant DNA Molecules*, National Insti-tutes of Health, Bethesda, Maryland, June 23, 1976.

24. "The Microsurgeon's Art Extends to Human Cells," *New Scientist*, May 21, 1970. This article reports on the work of Elaine Diacumakos and others. See E. G. Diacumakos, S. Holland, and P. Pecora, "A Microsurgical Methodology for Human Cells *in Vitro:* Evolution and Applications," *Proceedings of the National Academy of Sciences* 65 (1970):911–918. See Bibliography for subsequent Diacumakos papers.

25. K. W. Jeon, I. U. Lorch, and J. F. Danielli, "Reassembly of Living Cells from Dissociated Components," *Science*, March 20, 1970.

26. Henry Harris, *Cell Fusion* (Harvard University Press, 1970). Q. F. Ahkong, D. Fisher, W. Tampion, and J. A. Lucy, "Mechanisms of Cell Fusion," *Nature* 253 (1975):- 194–195. Lloyd C. Olson, "Methods of Cell Fusion with Germiston Virus," *Methods in Cell Biology* 14 (1976):11–22.

27. S. B. Carter, "Effects of Cytochalasins on Mammalian Cells," *Nature*, January 21, 1967.

28. Roger L. Ladda and Richard D. Estensen, "Introduction of a Heterologous Nucleus into Enucleated Cytoplasms of Cultured Mouse L-Cells," *Proceedings of the National Academy of Sciences*, November 1970. These researchers concluded from the results of their work that there was now good reason to believe that "the enucleation process may be used for utilization in nuclear transplantation studies in the mammalian egg similar to nuclear transplantation in amphibian eggs."

29. This was achieved by K. K. Sethi and H. Brandis at the Institute of Medical Microbiology and Immunology at the University of Bonn in West Germany. Writing in *Nature* ("Introduction of Mouse L Cell Nucleus into Heterologous Mammalian Cells," July 19, 1974), they reported using the fungal metabolite cytochalasin B to create viable hybrid cells, mixing the cellular parts of mice and men.

30. Dr. Christopher Graham of Oxford, for example, had used the Sendai virus to fuse mouse egg cells with mouse spleen and mouse bone-marrow cells, but he had not first enucleated his eggs and so had come up with hybrids containing double nuclei that did not get beyond the first cell division. (See Graham's "The Fusion of Cells with One- and Two-Cell Mouse Embryos," *Wistar Institute Symposium* 9 [1969]:19–33.) And Dr. Hilary Kaprowski of the Wistar Institute of the University of Pennsylvania, who had once proposed fusion as a means of cloning mammals, had made some progress in the direction of preparing mammalian eggs for fusion by using the enzyme pronase to strip them of their outer layer (zona pellucida). To keep them alive pending enucleation in the denuded state, he had found that lowering the temperature of the medium in which they were incubating was very effective. (See Symposium referred to above and "Closing in on Mammals," *Science News*, March 29, 1969.)

31. The "full treatment" as used here refers to what Gurdon called "serial transfer." In the first study cited by Darwin, Gurdon performed 461 nuclear transfers, of which 25 percent (124) were suitable for serial transfer. These 124 injected eggs formed partially cleaved blastulae; that is, embryos in which the transplanted nuclei underwent division in only one half of the egg, leaving the other half consisting only of cytoplasm. This came

216

about when the nuclei failed to divide during the egg's first cell division. Gurdon found it useful to take cells from the "normal" half of these abnormal embryos and then use *their* nuclei to "fertilize" freshly enucleated eggs, thus, in a sense, starting all over again. The advantage here was that these nuclei had enjoyed twice as much time as normal in which to replicate their DNA before undergoing their first division in the egg. Hence they were far more likely to be complete and capable of promoting normal growth. Because time did not permit Gurdon to attempt serial transfer with all 124 of his partial blastulae, he selected 14 of them, and from these 14 "full-treatment" nuclei he got six tadpoles, yielding, as Darwin looked at it, a 43 percent success rate. See J. B. Gurdon, R. A. Laskey, and O. R. Reeves, "The Developmental Capacity of Nuclei Transplanted from Keratinized Skin Cells of Adult Frogs," *Journal of Embryology and Experimental Morphology,* August 1975.

32. J. D. Bromhall, "Nuclear Transplantation in the Rabbit Egg," *Nature,* December 25, 1975. Dr. Bromhall (Oxford University and a former student of Dr. Gurdon's) used both microinjection and virally induced cell fusion to transfer body-cell nuclei into unfertilized rabbit eggs. The outer layers of the eggs used in the fusion experiments had been mechanically removed so that fusion could take place. The eggs were *not* enucleated prior to being fused or injected with rabbit karyoplasts; still, this was the most direct approach to mammalian cloning that had yet been described in print. Bromhall, noting that several previous investigators had fused mouse eggs with mouse body cells, hypothesized that the transplanted nuclei in these earlier experiments had failed to survive because no effort was made to synchronize egg and body cell. To overcome this obstacle to success he incubated some of his body-cell candidates in an atmosphere that contained nitrous oxide and enabled him to stop and start cell division at will. Thus was he able to achieve synchronization of cells. He could then microinject or fuse at the proper moment. Microsurgical manipulations of the eggs were performed at low temperatures (5 degrees C.). This "cold shock" served two purposes: first, it had been shown that eggs survive surgery better at low temperatures in this range, and second, the cold served to activate the egg cells 64 percent of the time. Cold shock *and* exposure to Sendai virus increased the activation rate to 83 percent. Overall, Bromhall reported finding cell fusion four times more likely to succeed than microsurgery; and synchronization of eggs and body cells doubled his success rate, whichever method was used. Bromhall was initially interested in seeing whether body-cell nuclei could be made to fuse with egg-cell nuclei. He established that this could occur and that cells thus fused could, in some cases, continue to divide at rates similar to those of eggs fertilized by sperm in vivo. He especially noted that a number of the eggs injected with body-cell nuclei underwent self-enucleation; that is, in order to accommodate the body-cell chromosomes, they cast out their own chromosomes (in what Bromhall called "pseudo polar bodies" similar to the chromosomes expelled by sperm-fertilized eggs). In a series of 155 eggs, 7.1 percent were found to have thus enucleated themselves. (Another 5.2 percent dispersed their chromosomes into their cytoplasms, where they were unlikely to contribute in any way to future development.) From the 7.1 percent, Bromhall got four embryos that cleaved regularly at in vivo rates all the way to the morula stage, at which point they might conceivably have been successfully implanted. Even if Bromhall counted the 5.2 percent as well as the 7.1 percent as part of the total "anucleate" eggs in this sample, he would have achieved, by my calculation, a success rate of about 20

percent (4 morulae from 19 anucleate candidates). The fact that seemingly viable embryos could result from these experiments at all, Bromhall concluded, "extends to the rabbit, and by inference to other mammals, the possibility of experiments which have so far been restricted to amphibians. . . . There is no apparent reason why a donor nucleus which can fuse with the egg nucleus after transplantation should not be able to support development by taking the place of the egg nucleus in an anucleate egg." Bromhall, I subsequently learned, was prevented from carrying his very fruitful inquiry any further when his funding, from the Cancer Research Campaign in Great Britain, ran out.

33. Audrey L. Muggleton-Harris and L. Hayflick, "Cellular Aging Studied by the Reconstruction of Replicating Cells from Nuclei and Cytoplasms Isolated from Normal Human Diploid Cells," *Experimental Cell Research* 103 (1976):321–330. "To our knowledge," these Stanford researchers concluded, "this is the first time that an isolated karyoplast and cytoplast of a normal human diploid cell have been reconstituted to form a viable, replicating cell."

34. For summary of Soviet research see Stephen Fulder, "Ginseng: Useless Root or Subtle Medicine," *New Scientist,* January 20, 1977.

35. R. H. Yonemoto, P. B. Chretien, and T. F. Fehniger, "Enhanced Lymphocyte Blastogenesis by Oral Ascorbic Acid," *American Society of Clinical Oncologists,* 1976, p. 288. See also R. Hume and E. Weyers, "Changes in Leucocyte Ascorbic Acid During the Common Cold," *Scottish Medical Journal* 18 (1973):3–7.

36. Darwin was particularly impressed by the work of Drs. D. J. MacRae and S. M. Mohamedally, both of Mother's Hospital in London. They had reported (in "Effect of Abdominal Decompression on the Metabolism of the Fetoplacental Unit," *Proceedings of the Royal Society of Medicine,* May 1970) on some of the finer metabolic effects of decompression, specifically examining the effects of the technique on the output of the maternal hormones estriol and pregnanediol. These hormones are crucial for the maintenance of pregnancy and were chosen by the London researchers because they are such good indicators of placental and overall fetal health. If placental function was good, then these hormones would be measurable within a certain, well-defined range; if function was poor, the levels would fall below this range. MacRae and associates decided that, if decompression was truly useful in overcoming or preventing placental insufficiency, then it should have a direct, boosting effect on these important hormones. They took seventeen pregnant women in whom these hormones were at depressed levels and treated them with decompression. "A rise in hormone levels was obtained in fifteen of the seventeen cases treated," the researchers reported. The rise in each case, moreover, coincided with the onset of decompression treatments. Most of these women had long histories of miscarriage but were able to carry babies to term for the first time with decompression. Each decompression session lasted half an hour. There were three of these a week in early pregnancy and two a day later on. One of the two failures, it was noted, was able to take only two treatments per week throughout pregnancy "because of domestic difficulties." A subsequent report, also of great interest to Darwin, appeared in the *Journal of Obstetrics and Gynaecology of the British Commonwealth,* January 1971. It was written by Alan Coxon, an obstetrician, and J. W. Haggith, a physicist. In this report ("The Effects of Abdominal Decompression on Vascular Hemodynamics in Pregnancy and Labor"), the Newcastle researchers demonstrated reproducible blood-flow increase into the placenta during decompression. These increases were observable even during labor, when oxygen supply

218

to the baby is particularly imperiled. Decompression, they said, produced a 30 percent increase in blood supply when there were no contractions, a 15 percent increase when there were.

37. "First Documented Pregnancy from *in-Vitro* Fertilization Is Reported," *Ob.-Gyn. News*, June 15, 1976.

38. "Bouncing Baby Baboon Has Two Mothers," *Medical World News*, July 17, 1976.

39. L. J. Quinn and R. A. McKeown, "Abdominal Decompression During the First State of Labor," *American Journal of Obstetrics and Gynecology*, 83 (1962):458. Also L. J. Quinn, P. Dorr, and R. Bruyere, "Experiences with Abdominal Decompression During Labor," *Journal of Obstetrics and Gynaecology of the British Commonwealth*, 71 (1964):- 934.

40. I was put in mind of this category of individual when I read an article by British scientist Sir Peter Medawar in an October 1977 issue of *New York Review of Books* called "Fear and DNA." Sir Peter, it seems, perceives it his duty to apologize for some of the new recombinant DNA work (the article is supposed to be a review of three recent books on this research). It seems that he does not, however, want to be perceived by others as too ardent an apologist, sensing perhaps that if the new work were all benign it would require no apology. What I find disturbing arises out of Sir Peter's apparent need to demonstrate that, even though he is basically in the camp of those who want to forge ahead with this work, he is still a thoughtful person mindful of some of the risks inherent in genetic tinkering. This need, unfortunately, finds its fulfillment in a diatribe against human cloning which is nothing other than a protracted non sequitur. Apropos of nothing, he suddenly recalls that he once had a conversation with noted author and scientist Dr. Jacques Monod, at that time director of the Pasteur Institute in France. Medawar acknowledges that Monod "was as well qualified as anybody in the world to express an authoritative opinion" on matters related to genetic engineering. Monod, he states, believed that human cloning was "a definite possibility" in the near future. But then, without any further word on Monod's reasons for believing this, Medawar proceeds to state that *he* believes that "the enterprise is simply not on," that it "would require a degree of organization that would make the mobilization and deployment of an army seem like running a Sunday school picnic." To this he sees fit to add that only the "foolish" or "simpletons" would become involved in such a "scheme" in the first place. Elsewhere he adds to these epithets the adjective "idiotic." Sir Peter demonstrates himself out of touch with the views of several of his informed peers, including Monod, and demonstrates as well that he hasn't been keeping abreast of the published data, which clearly indicate that man is at least within a hair's breadth of being able to clone himself—provided he wants to badly enough. Indeed, Sir Peter seems to have forgotten that, as the very first sentence of his article, he has written, "It is the great glory as it is also the great threat of science that everything which is in principle possible can be done if the intention to do it is sufficiently resolute." What bothers me most is the extreme language he uses in his effort to condemn human cloning. He protests too much and too gracelessly; I come to suspect him of a basic insincerity. It seems as if he is saying, "You see how trustworthy I am when I tell you this new DNA work is all right because I'm four-square against the mad-scientist stuff, like cloning."

I'm left wondering why Sir Peter, instead of devoting so much space to human cloning in an article that purports to be a review of the new DNA work, hasn't focused instead on some of the real dangers of genetic engineering. Even while supporting most of the work (as I do), he might, for example, have examined the NIH *Guidelines* to see whether there were holes in them that need plugging in the public interest. By acknowledging some of the shortcomings as well as noting the strengths of the DNA work, he might better (than through the deprecation of cloning) enhance his credibility as an apologist for this research. For when it comes to holes in the *Guidelines* there is, alas, something worth writing about. Sir Peter might have mentioned, for example, the fears of some personnel working on the home grounds of NIH in Bethesda. The radiation workers—of all people—are worried about their own personal safety and have been since the biologists set up shop right next door to them. The experiment that concerns some of them has to do with a plan to transplant some of the genes of the cancer-causing *Polyoma* virus into *E. coli* bacteria which will subsequently be fed to mice. *Polyoma* causes cancer in rodents. NIH *Guidelines* specify that when you experiment with cancer-causing viruses you must use a weakened strain of *E. coli*, one that cannot, we are told, survive in human or other animal intestines. Obviously, however, in order for this experiment to work, the attenuated version of the bacterium will not do, for the plan is to put the *Polyoma*-laced *E. coli* into the intestines of mice, in an effort to figure out which of the viral genes are responsible for starting cancer. The requirement to use the enfeebled *E. coli* has, therefore, been waived.

The radiation people, who have certainly had experience in dealing with dangerous substances, are not reassured by the containment facility in which the *Polyoma* work is to be performed. They know how easy it is, through the smallest bit of negligence, for something to go amiss. They are worried, some of them at least, that some of the cancer-carrying *E. coli* might get out of its "cage" and into *their* intestines. From that point on it might spread very rapidly through the entire population—into *your* intestines and mine, with what effect no one can say; but some find it more than merely "interesting" that Dr. Sarah E. Stewart, codiscoverer of the *Polyoma* virus in 1953, recently died of cancer and that several individuals with whom she lived also have the disease. This work (see Judith Randal, "Life from the Labs: Who Will Control the New Technology?" *The Progressive*, March 1977) is either going on now or still scheduled to go forward.

Meanwhile, *Science* magazine has acknowledged that there have already been what it terms some "narrow escapes" in the new research. (See Nicholas Wade, "Dicing with Nature," *Science* 195 [1977]:378.) One of these was able to occur, apparently, because the *Guidelines* fail to cover the in vivo creation of potentially deadly new life forms, focusing instead only on in vitro (in the test tube) products of gene-splicing. Dr. A. Chakrabarty of the General Electric Research and Development Center in Schenectady, New York, created an *E. coli* bacterium with a new gene using plasmid engineering. This new gene coded for the production of cellulase, an enzyme that breaks down cullulose, a plant protein that is normally indigestible by humans. (It is this sort of thing that could lead to our being able to eat hay or grass.) Everything seemed to be going along all right until Dr. Chakrabarty suddenly realized that breakdown products of cullulose might be imperfectly absorbed in the lower intestine, resulting in gas buildup and perpetual stomach upset. There was more to this than just breaking down cullulose. An *E. coli* that was only halfway capable of processing plant proteins might be a very dangerous bug. "Should such an *E. coli* gain a selective advantage," *Science* noted, "and spread throughout the population, the result might be a large number of people suffering from chronic, maybe fatal, diarrhea." And because *E. coli*, which continues to be the primary vehicle of all this recombinant work, has coexisted with us since the beginning of time, apparently, and is everywhere, it would be virtually impossible to rid ourselves of this gaseous man-made plague should it ever gain a foothold. Chakrabarty destroyed his new life form. There may be others, however, who will not be so quick to realize the dangers of their work.

BIBLIOGRAPHY

Unsigned Reports

"Bouncing Baby Baboon Has Two Mothers," *Medical World News*, July 17, 1976.

"Clonal Reproduction: Closing In on Mammals," *Science News*, March 29, 1969.

"Cloned Cauliflowers for Greater Productivity," *New Scientist*, June 18, 1970.

"Cloning: The Ethical Question," *Science Digest*, August 1971.

"Doomsday: Tinkering with Life," *Time*, April 18, 1977.

"Fertilized Ova Transferred Across the Atlantic," *Science Journal*, November 1970.

"First Documented Pregnancy from *in-Vitro* Fertilization Is Reported," *Ob.-Gyn. News*, June 15, 1976.

Guidelines for Research Involving Recombinant DNA Molecules, National Institutes of Health, Bethesda, Maryland, June 23, 1976.

"The Microsurgeon's Art Extends to Human Cells," *New Scientist*, May 21, 1970

Molecular Cloning: Powerful Tool for Studying Genes," *Science* 191 (1976):-1160–1162.

"Mouse Embryos Grow in Absence of Uterine Tissue," *Ob. Gyn. News*, May 15, 1971.

"Protection of Human Subjects–Policies and Procedures," *Federal Register* (National Institutes of Health) 38 (November 16, 1973): 31738–31749.

"[Italian] Reports Human Egg Fertilized in Test Tube," *New York Post*, January 13, 1961.

"Sci-fi on the Range," *Newsweek*, September 4, 1972.

"Steward on the Cloning of Plants—and the Possibility of Cloning People," *New Scientist,* May 22, 1969.

"Theoretical Ferment in Embryology," *Science News,* February 28, 1970.

"Tree Clones," *Science Digest,* April 1975.

"Work with Reimplanting Ova Moves Another Step Ahead," *Journal of the American Medical Association,* January 4, 1971.

Signed Reports and Books

Ahkong, Q. F., Fisher, D., Tampion, W., and Lucy, J. A., "Mechanisms of Cell Fusion," *Nature* 253 (1975):194–195.

Austin, C. R., "Capacitation of Sperm," *New Scientist,* July 31, 1969.

Balfour-Lynn, S., "Parthenogenesis in Human Beings," *Lancet* 1 (1956):1071–1072.

Barnes, D., Tuffrey, M., and Graham, C. F., "Reduced Levels of Serum Haemolytic Complement and Renal Lesions in Ovum-Fusion-Derived Tetraparental Mouse Chimeras," *Scandinavian Journal of Immunology* 3 (1974):789–796.

Barnes, R. D., "In-Vitro Fertilization" (letter), *Lancet* 1 (1976):1016–1017.

Beckwith, J., "Gene Expression in Bacteria and Some Concerns About the Misuse of Sciences," *Bacteriological Review* 34 (1970):224.

Bennett, W., and Gurin, J., "Science that Frightens Scientists: The Great DNA Debate," *Atlantic Monthly,* February 1977.

Berg, P., "Seeking Wisdom in Recombinant DNA Research," *Federation Proceedings* (Federation of American Society for Experimental Biology) 35 (1976):1133–1135.

Blecher, J. A., "Cardio-Pulmonary Physiology of Abdominal Decompression," *Lancet* 2 (1967):614.

Bolund, L., Ringertz, N. R., and Harris, H., "Changes in the Cytochemical Properties of Erythrocyte Nuclei Reactivated by Cell Fusion," *Journal of Cell Science* 4 (1969):71–87.

Bonner, J., "Beyond Man's Genetic Lottery," in *The Next Billion Years: Mankind's Future in a Cosmic Perspective.* Moffett Field, Calif.: Ames Research Center, NASA, 1974.

Briggs, R., and King, T. J., "Transplantation of Living Nuclei from Blastula Cells into Enucleated Frogs' Eggs," *Proceedings of the National Academy of Sciences* 38 (1952):455–463.

Bromhall, J. D., *An Investigation of Nuclear Transplantation in the Mammalian Egg.* D.Phil. thesis, Oxford University, 1975.

———, "Nuclear Transplantation in the Rabbit Egg," *Nature* 258 (1975):-719–722.

———, personal communication, July 18, 1977.

Brothers, A. J., "Stable Nuclear Activation Dependent on a Protein Synthesized During Oogenesis," *Nature* 260 (1976):112–115.

224

Bulmer, M. G., *The Biology of Twinning in Man.* New York and London: Hawthorn Books, 1972.

Burton, B. K., Gerbie, A. B., and Nadler, H. L., "Present Status of Intrauterine Diagnosis of Genetic Defects," *American Journal of Obstetrics and Gynecology* 118 (1974):718–746.

Callahan, D., "New Beginnings in Life: A Philosopher's Response," in Michael P. Hamilton, ed., *The New Genetics and the Future of Man.* Grand Rapids, Mich.: William B. Eerdmans Publishing Company, 1972.

Carrel, A., *Man, the Unknown.* New York: Harper & Brothers, 1939.

Carter, S. B., "Effects of Cytochalasins on Mammalian Cells," *Nature* 213 (1967):261–264.

Cavalieri, L., "New Strains of Life—or Death," *New York Times Magazine,* August 22, 1976.

Chakrabarty, A. M., "Which Way Genetic Engineering?" *Industrial Research,* January 1976.

Chang, M. C., "Fertilization and Normal Development of Follicular Oocytes in Rabbit," *Science* 121 (1955):867–869.

Chargaff, E., "Building the Tower of Babble," *Nature* 248 (1974):776–782.

——, "A Slap at the Bishops of Asilomar," *Science* 190 (1975):135.

——, and Simring, F. R., "On the Dangers of Genetic Meddling" (letter), *Science* 192(1976):983.

Cheng, T., and Voqui, T. H., "Regeneration of Douglas Fir Plantlets Through Tissue Culture," *Science* 198 (1977):306–307.

Cohen, S. N., "The Manipulation of Genes," *Scientific American,* July 1975.

——, "Recombinant DNA: Fact and Fiction," *Science* 195 (1977):654–657.

Coxon, A., "The Effects of Abdominal Decompression on Vascular Hemodynamics in Pregnancy and Labor," *Journal of Obstetrics and Gynaecology of the British Commonwealth,* January 1971, pp. 49–54.

Crick, F., *Of Molecules and Men.* Seattle and London: University of Washington Press, 1966.

Cuellar, O., "Intraclonal Histocompatability in a Parthenogenetic Lizard: Evidence of Genetic Homogeneity," *Science* 193 (1976):150–153.

Danielli, J. F., "Artificial Synthesis of New Life Forms," *Bulletin of the Atomic Scientists* 28 (December 1972):20–24.

——, "Industry, Society and Genetic Engineering," *The Hastings Center Report,* December 1972.

——, Lorch, I. J., Ord, M. J., and Wilson, E. G., "Nucleus and Cytoplasm in Cellular Inheritance," *Nature* 176 (1955):1114–1115.

Davis, B. D., "Prospects for Genetic Intervention in Man," *Science* 170 (1970):1279–1283.

——, "Genetic Engineering: How Great the Danger?" *Science* 186 (1974):-309–325.

Davis, B. N., "Darwin, Pasteur and the Andromeda Strain," in *DNA Recombinant Molecule Research, Supplemental Report II,* prepared for the Subcommittee on Science, Research and Technology of the Committee on

Science and Technology, U.S. House of Representatives, December 1976.

Deak, I., Sidebottom, E., and Harris, H., "Further Experiments on the Role of the Nucleolus in the Expression of Structural Genes," *Journal of Cell Science* 11 (1972):379–391.

Diacumakos, E. G., "Methods for Microsurgical Production of Mammalian Somatic Cell Hybrids and Their Analysis and Cloning," *Methods in Cell Biology* 10 (1975):147–156.

―――, Holland, S., and Pecora, P., "A Microsurgical Methodology for Human Cells *in Vitro:* Evolution and Applications," *Proceedings of the National Academy of Sciences* 65 (1970):911–918.

―――, ―――, and ―――, "Microsurgical Studies on Human Cells and Cloning of HeLa Cells," *Nature* 232 (1971):28–32.

―――, ―――, and ―――, "Chromosome Displacement in and Extraction from Human Cells at Different Mitotic Stages," *Nature* 232 (1971):33–36.

―――, and Tatum, E. L., "Fusion of Mammalian Somatic Cells by Microsurgery," *Proceedings of the National Academy of Sciences* 69 (1972):2959–2962.

Diamante, A. O., "Ten *in-Vitro* Implants in Humans 'Failures,'" *Ob.Gyn. News,* January 15, 1974.

DiBerardino, M. A., and Hoffner, N., "Development and Chromosomal Constitution of Nuclear-Transplants Derived from Male Germ Cells," *Journal of Experimental Zoology* 176 (1971):61–72.

Dobzhansky, T., "Changing Man," *Science* 155 (1967):409–415.

Dolezal, A., "Increase in Uterine Blood Flow in the First Stage of Labor During Abdominal Decompression as Measured with ^{131}I-HSA," *American Journal of Obstetrics and Gynecology,* January 1970.

Dupey-Coin, A. M., Ege, T., Bouteille, M., and Ringertz, N. R., "Ultrastructure of Chick Erythrocyte Nuclei Undergoing Reactivation in Heterokaryons and Enucleated Cells," *Experimental Cell Research* 101 (1976):355–369.

Edwards, R. G., "Studies on Human Conception," *American Journal of Obstetrics and Gynecology* 117 (1973):587–601.

―――, and Fowler, R. E., "Human Embryos in the Laboratory," *Scientific American* 223 (1970):44–54.

―――, Steptoe, P. C., and Purdy, J. M., "Fertilization and Cleavage *in Vitro* of Preovulatory Human Oocytes," *Nature* 227 (1970):1307–1309.

Ettinger, R., *Man into Superman.* New York: St. Martin's Press, 1972.

Etzioni, A., *Genetic Fix.* New York: Macmillan Publishing Company, 1973.

Feinberg, G., *The Prometheus Project: Mankind's Search for Long Range Goals.* Garden City, N.Y.: Doubleday & Company, 1969.

Fletcher, J., "Ethical Aspects of Genetic Controls," *New England Journal of Medicine* 285 (1971):776–783.

―――, "New Beginnings in Life: A Theologian's Response," in Michael P. Hamilton, ed., *The New Genetics and the Future of Man,* Grand Rapids, Mich.: William B. Eerdmans Publishing Company, 1972.

―――, *The Ethics of Genetic Control.* Garden City, N.Y.: Anchor

226

Press/Doubleday & Company, 1974.

Francouer, R. T., *Utopian Motherhood.* Garden City, N.Y.: Doubleday & Company, 1972.

Frisch, B., "Genetics: What It Will Do for the Next Generation," *Science Digest,* March 1967.

Gaddis, V., and Gaddis, M., *The Curious World of Twins.* New York: Hawthorn Books, 1972.

Galston, A. W., "Here Come the Clones," *Natural History,* February 1975.

Gaylin, W., "Frankenstein Myth Becomes Reality," *New York Times Magazine,* March 5, 1972.

Gedda, L., *Twins in History and Science.* Springfield, Ill.: Charles C. Thomas, Publishers, 1961.

Glass, H. B., "Genetics and Surgery," *Bulletin of the American College of Surgeons,* November 1972.

Gould, G. M., and Pyle, W. L., *Anomalies and Curiosities of Medicine.* New York: Sydenham Publishers, 1937.

Graham, C. F., "The Fusion of Cells with One- and Two-Cell Mouse Embryos," *Wistar Institute Symposium* 9 (1969):19–33.

———, "Parthenogenetic Mouse Blastocysts," *Nature* 226 (1970):165–167.

———, "Virus Assisted Fusion of Embryonic Cells," *Acta Endocrinologica* [*supplement*] (Copenhagen) 153 (1971):154–167.

———, "The Necessary Conditions for Gene Expression During Early Mammalian Development," *Symposium of the Society of Developmental Biologists* 31 (1973):201–224.

———, "The Production of Parthenogenetic Mammalian Embryos and Their Use in Biological Research," *Biological Review* 49 (1974):399–424.

Grobstein, C., "Recombinant DNA Research: Beyond the NIH *Guidelines,*" *Science* 194 (1976):1133–1135.

Guentert, K., "Will Your Grandchild Be a Test-Tube Baby?" *U.S. Catholic,* June 1977.

Gurdon, J. B., personal communication, September 20, 1968.

———, "Transplanted Nuclei and Cell Differentiation," *Scientific American,* December 1968.

———, *The Control of Gene Expression in Animal Development.* Cambridge: Harvard University Press, 1974.

———, "Molecular Biology in a Living Cell," *Nature* 248 (1974):772–776.

———, personal communication, May 13, 1977.

———, and Laskey, R. A., "The Transplantation of Nuclei from Single Cultured Cells into Enucleate Frogs' Eggs," *Journal of Embryology and Experimental Morphology* 24 (1970):499–526.

———, DeRobertis, E. M., and Partington, G., "Injected Nuclei in Frog Oocytes Provide a Living Cell System for the Study of Transcriptional Control," *Nature* 180 (1976):116–120.

———, Laskey, R. A., and Reeves, O. R., "The Developmental Capacity of Nuclei Transplanted from Keratinized Skin Cells of Adult Frogs," *Journal*

227

of *Embryology and Experimental Morphology* 34 (1975):93–112.

Halacy, D. S., Jr., *Genetic Engineering: Threat or Promise?* New York: Harper & Row Publishers, 1974.

Haldane, J. B. S., "Biological Possibilities in the Next Ten Thousand Years," in Gordon Wolstenholme, ed., *Man and His Future*. Boston: Little, Brown & Company, 1963.

Hamilton, W. J., and Glenister, T. W., "Human Life in the Test Tube," *The Times of London* (letter), February 19, 1969.

Harris, H., Sidebottom, E., Grace, D. M., and Bramwell, M. "The Expression of Genetic Information: A Study with Hybrid Animal Cells," *Journal of Cell Science* 4 (1969):499–526.

Hennen, S., "Influence of Spermine and Reduced Temperature on the Ability of Transplanted Nuclei to Promote Normal Development in Eggs of *Rana pipiens*," *Proceedings of the National Academy of Sciences* 66 (1970):-630–637.

Heyns, O. S., "The Present Range of Value of Abdominal Decompression," *Proceedings of the Royal Society of Medicine* 55 (1962):459–461.

———, *Abdominal Decompression*. Johannesburg: Witwatersrand University Press, 1963.

———, and Blecher, J. A., "Abdominal Decompression in the Treatment of Toxemias of Pregnancy," *Lancet* 2 (1967):621.

———, Samson, J. M., and Graham, J. A. C., "Influence of Abdominal Decompression on Intra-Amniotic Pressure and Fetal Oxygenation," *Lancet* 1 (1962):289–292.

Hirschhorn, K., "On Redoing Man," *Commonweal*, May 17, 1968.

Howard, T., and Rifkin, J., *Who Should Play God?* New York: Delacorte Press, 1977.

Hurock, G. A., "Gene Therapy and Genetic Engineering: Frankenstein Is Still a Myth, But It Should Be Read Periodically," *Indiana Law Journal* 48 (Summer, 1973):544.

Huxley, A., *Brave New World*. New York: Harper & Brothers, 1946.

Illmensee, K., and Mintz, B., "Totipotency and Normal Differentiation of Single Teratocarcinoma Cells Cloned by Injection into Blastocysts," *Proceedings of the National Academy of Sciences* 73 (1976):549–553.

Jeon, K. W., Lorch, I. J., and Danielli, J. F., "Reassembly of Living Cells from Dissociated Components," *Science* 167 (1970):1626–1627.

Judson, H. F., "Fearful of Science," *Harper's*, March 1975.

Karp, L. E., *Genetic Engineering: Threat or Promise?* Chicago: Nelson-Hall, 1976.

Kass, L. R., "Babies by Means of *in-Vitro* Fertilization: Unethical Experiments on the Unborn?" *New England Journal of Medicine* 285 (1971):1174–1179.

———, "New Beginnings in Life," in Michael P. Hamilton, ed., *The New Genetics and the Future of Man*. Grand Rapids, Mich.: William B. Eerdmans Publishing Company, 1972.

228

Kaufman, H. H., Huberman, E., and Sachs, L., "Genetic Control of Haploid Parthenogenetic Development in Mammalian Embryos," *Nature* 254(1975): 694–695.

Kennedy, E., "Baby Bubble Sequel," *Ladies' Home Journal*, October, 1971.

Kiewit, F., "Embryo Transplanting May Alter Cattle Industry," *Kansas City Star*, August 1, 1971.

King, T. J., and Briggs, R., "Serial Transplantation of Embryonic Nuclei," *Cold Spring Harbor Symposium* 21 (1956):271–290.

———, and DiBerardino, M. A., "Transplantation of Nuclei from the Frog Renal Adenocarcinoma—Development of Tumor Nuclear Transplant Embryos," *Annals of the New York Academy of Science* 126 (1965):115–126.

———, and McKinnel, R. G., "An Attempt to Determine the Developmental Potentialities of the Cancer Cell Nucleus by Means of Transplantation," in *Cell Physiology of Neoplasia.* University of Texas Press, 1960.

Ladda, R. L., and Estensen, R. D., "Introduction of a Heterologous Nucleus into Enucleated Cytoplasms of Cultured Mouse L-Cells," *Proceedings of the National Academy of Sciences* 67 (1970):1528–1533.

Lappé, M., "The Human Uses of Molecular Genetics," *Federation Proceedings* (Federation of the American Society for Experimental Biology) 34 (1975):-1425–1427.

———, "Realities of 'Genetic Engineering,' " *Medical Research Engineering* 12 (1976):25–29.

Laskey, R. A., and Gurdon, J. B., "Genetic Content of Adult Somatic Cells Tested by Nuclear Transplantation from Cultured Cells," *Nature* 228 (1970):1332–1334.

Lederberg, J., "Experimental Genetics and Human Evolution," *Bulletin of the Atomic Scientists* 23 (October 1966):4–11.

———, "Orthobiosis: The Perfection of Man," in A. Tiselius and S. Nilsson, eds., *The Place of Value in a World of Fact.* New York and London: John Wiley & Sons, 1971.

———, "Genetic Engineering, or the Amelioration of Genetic Defect," *Pharos* 34 (1971):9–12.

———, "DNA Splicing: Will Fear Rob Us of Its Benefits?" *Prism*, November 1975.

———, "Law and Cloning—The State as Regulator of Gene Function," in A. Milunsky and G. J. Annas, eds., *Genetics and the Law.* New York: Plenum Press, 1976.

Lessing, L., "Into the Core of Life Itself," *Fortune*, March 1966.

Lindeman, B., "The Twins Who Found Each Other," *Saturday Evening Post*, March 21, 1964.

Lovlie, A., "Signal for Cell Fusion," *Nature* 263 (1976):779–81.

Lowall, G., "Killer-on-the-Loose: The Sinister Shadow of 'XYY,' " *Argosy*, February 1968.

Luria, S., "Modern Biology: A Terrifying Power," *The Nation*, October 20, 1969.

229

MacRae, D. J., "Effect of Abdominal Decompression on the Metabolism of the Fetoplacental Unit," *Proceedings of the Royal Society of Medicine* 63 (1970):502–505.

Marx, J. L., "Embryology: Out of the Womb—into the Test Tube," *Science* 182 (1973):811–814.

Matthews, D. D., and Loeffler, F. E., "The Effect of Abdominal Decompression on Fetal Oxygenation During Pregnancy and Early Labor," *Journal of Obstetrics and Gynaecology of the British Commonwealth* 75 (1968):268.

Medawar, P. B., "Fear and DNA," *New York Review of Books*, October 27, 1977.

———, and Medawar, J. S., "Revising the Facts of Life," *Harper's*, February 1977.

Mintz, B., "Formation of Genetically Mosaic Mouse Embryos and Early Development of Lethal (t^{12}/t^{12})-Normal Mosaics," *Journal of Experimental Zoology* 157 (1964):267–271.

———, and Illmensee, K., "Normal Genetically Mosaic Mice Produced from Malignant Teratocarcinoma Cells," *Proceedings of the National Academy of Sciences* 72 (1975):3585–3589.

Monod, J., *Chance and Necessity.* New York: Alfred A. Knopf, 1971.

Morrow, J. G., "The Prospects for Gene Therapy in Humans," *Annals of the New York Academy of Science* 265 (1976):13–21.

Muggleton-Harris, A. L., and Hayflick, L., "Cellular Aging Studied by the Reconstruction of Replicating Cells from Nuclei and Cytoplasms Isolated from Normal Human Diploid Cells," *Experimental Cell Research* 103 (1976):321–330.

———, and Pezzella, K., "The Ability of the Lens Cell Nucleus to Promote Complete Embryonic Development and Its Applications to Opthalmic Gerontology," *Experimental Gerontology* 7 (1972):427–431.

Mukherjee, A. B., and Cohen, M. M., "Viable Mouse Offspring from Blastocysts Cultured *in Vitro*," *Nature* 228 (1970):472.

Muller, H. J., "Our Load of Mutations," *American Journal of Human Genetics* 2 (1950):111–176.

———, "The Guidance of Human Evolution," *Perspectives in Biology and Medicine* 3 (1959):1–43.

———, "Human Evolution by Voluntary Choice of Germ Plasm," *Science* 134 (1961):643–649.

———, *Studies in Genetics: The Selected Papers of H. J. Muller.* Bloomington: Indiana University Press, 1962.

Munson, R., ed., *Man and Nature: Philosophical Issues in Biology.* New York: Delta Books/Dell Publishing Company, 1971.

Nadler, H. L., "Prenatal Diagnosis of Inborn Defects: A Status Report," *Hospital Practice* 10 (No. 6, 1975):41–51.

Newman, H. H., Freeman, F. N., and Holzinger, K. J., *Twins: A Study of Heredity and Environment.* Chicago: University of Chicago Press, 1966.

Olson, L. C., "Methods of Cell Fusion with Germiston Virus," *Meth-*

ods in Cell Biology 14 (1976):11–22.

Packard, V., *The People Shapers.* Boston: Little, Brown & Company, 1977.

Packer, P., *Death of the Other Self.* New York: Cowles Book Company, 1970.

Penrose, L. S., "Propagation of the Unfit," *Lancet* 2 (1950):425–427.

Pincus, G., and Shapiro, H., "Further Studies on the Parthenogenetic Activation of Rabbit Eggs," *Proceedings of the National Academy of Sciences* 26 (1940):163–165.

Poste, G., Alexander, D., and Reeve, P., "Enhancement of Virus-Induced Cell Fusion by Phytohemagglutinin," *Methods in Cell Biology* 14 (1976):1–10.

Quinn, L. J., and McKeown, R. A., "Abdominal Decompression During the First Stage of Labor," *American Journal of Obstetrics and Gynecology* 83 (1962):458.

———, Dorr, P., and Bruyere, R., "Experiences with Abdominal Decompression During Labor," *Journal of Obstetrics and Gynaecology of the British Commonwealth* 71 (1964):934.

Ramsey, P., *Fabricated Man.* New Haven: Yale University Press, 1970.

Randal, J., "Life from the Labs: Who Will Control the New Technology?" *The Progressive*, March 1977.

Range, G., " 'Surrogate Mother' Is Recruited by Ad for Artificial Insemination," *Ob.Gyn. News*, July 1, 1976.

Rifkin, J., Gordon, L., and Smith, D., "DNA," *Mother Jones*, February–March 1977.

Rivers, C., "Cloning: A Generation Made to Order," *Ms.*, June 1976.

Rock, J., and Hertig, A. T., "The Human Conceptus During the First Two Weeks of Gestation," *American Journal of Obstetrics and Gynecology* 55 (1948):6.

———, and Menkin, M. F., *"In-Vitro* Fertilization and Cleavage of Human Ovarian Eggs," *Science* 100 (1944):105–107.

Rogers, M., "The Pandora's Box Congress," *Rolling Stone*, June 19, 1975.

Rorvik, D. M., "Preview: The Test-Tube Generation," *Esquire*, April 1969.

———, "Artificial Inovulation," *McCall's*, May 1969.

———, "Surgery on the Unborn," *Look*, November 4, 1969.

———, "Asexual Propagation," *Science Digest*, December 1969.

———, "Taking Life in Our Own Hands," *Look*, May 18, 1971.

———, *Brave New Baby: Promise and Peril of the Biological Revolution.* Garden City, N.Y.: Doubleday & Company, 1971.

———, *As Man Becomes Machine: Evolution of the Cyborg.* Garden City, N.Y.: Doubleday & Company, 1971.

———, "Mechanizing the Mind for Good and Evil," *Catholic Digest*, March 1972.

———, "Behavior Control: Big Brother Comes," *Intellectual Digest*, January 1974.

———, "Bringing the War Home," *Playboy*, September 1974.

———, "The Embryo Sweepstakes," *New York Times Magazine*, September 15, 1974.

————, *Good Housekeeping Woman's Medical Guide.* New York: Hearst Books, 1974.

————, "Embryo Transplants," *Good Housekeeping,* June 1975.

————, "The Gender Enforcers," *Rolling Stone,* October 9, 1975.

————, and Heyns, O. S., *Decompression Babies.* New York: Dodd, Mead & Company, 1973.

————, and Shettles, L. B., *Your Baby's Sex: Now You Can Choose.* New York: Dodd, Mead & Company, 1971.

————, and ————, *Choose Your Baby's Sex.* New York: Dodd, Mead & Company, 1977.

Rosenfeld, A., *The Second Genesis.* New York: Prentice-Hall, 1969.

————, "Should We Tamper with Heredity?" *Saturday Review,* July 26, 1975.

Rostand, J., *Can Man Be Modified?* New York: Basic Books, 1959.

————, *Humanly Possible.* New York: Saturday Review Press, 1973.

Rugh, R., and Shettles, L. B., *From Conception to Birth: The Drama of Life's Beginnings.* New York: Harper & Row Publishers, 1971.

Scheinfeld, A., *Twins and Supertwins.* Philadelphia and New York: J. B. Lippincott Company, 1967.

Schopf, J., "Evolution of Earth Biosphere," in *The Next Billion Years: Mankind's Future in a Cosmic Perspective.* Moffett Field, Calif.: Ames Research Center, NASA, 1974.

Schumacher, G. F. B., Brackett, B. G., Fletcher, J., Marik, J. J., Mastroianni, L., Jr., Shettles, L. B., Tejada, R., and Taylor, E. T., *"In-Vitro* Fertilization of Human Ova and Blastocyst Transfer: An Invitational Symposium," *The Journal of Reproductive Medicine* 11 (November 1973):192–204.

Sethi, K. K., and Brandis, H., "Introduction of Mouse L Cell Nucleus into Heterologous Mammalian Cells," *Nature* 250 (1974):225–226.

Shettles, L. B., "A Morula Stage of Human Ovum Developed *in Vitro,"* *Fertility and Sterility* 6 (1955):287–289.

————, *Ovum Humanum.* Munich and Berlin: Urban & Schwarzenberg, 1960.

————, "Human Fertilization," *Obstetrics and Gynecology* 20 (1962):750–754.

————, "Einbettung und Ansieldlung des menshclichen Embryos," *Medizinische Klinik* 58 (1963):1623–1624.

————, "In the Beginning," *Bulletin of the Sloane Hospital for Women,* vol. X, Winter 1964, pp. 246–261.

————, "Human Fertilization and Development from the Inner Cell Mass," in E. Philipp, ed., *The Scientific Foundations of Obstetrics and Gynecology.* Philadelphia: F. A. Davis Company, 1970.

————, personal communications, 1970–1977.

————, "Human Blastocyst Grown *in Vitro* in Ovulation Cervical Mucus," *Nature* 229 (1971):343.

————, "Use of the Y Chromosome in Prenatal Sex Determination," *Nature* 230 (1971):52–53.

Shields, J., *Monozygotic Twins Brought Up Apart and Brought Up Together: An Investigation into the Genetic and Environmental Causes of Variation in*

Personality. London: Oxford University Press, 1962.

Singer, M. F., "The Recombinant DNA Debate," *Science* 196 (1977):127.

Sinsheimer, R., "Genetic Engineering: The Modification of Man," *Impact of Science on Society* 20 (1970):279–291.

———, "Troubled Dawn for Genetic Engineering," *New Scientist* 68 (1975):-148–151.

———, "On Coupling Inquiry and Wisdom," *Federation Proceedings* (Federation of American Society for Experimental Biology) 35 (1976):2540–2542.

Stent, G. S., "Molecular Biology and Metaphysics," *Nature* 248 (1974):779–781.

Steptoe, P. C., personal communications, February 11, 1974, and February 26, 1974.

———, and Edwards, R. G., "Laparascopic Recovery of Preovulatory Human Oocytes after Priming of Ovaries with Gonadotrophins," *Lancet* 1 (1970):-683–689.

———, and ———, "Reimplantation of a Human Embryo with Subsequent Tubal Pregnancy," *Lancet* 1 (1976):880–882.

Steward, F. C., "From Cultured Cells to Whole Plants: The Induction and Control of Their Growth and Differentiation," *Proceedings of the Royal Society [Biology]* 175 (1970):1–30.

———, "Totipotency, Variation and Clonal Development of Cultured Cells," *Endeavor,* September 1970.

———, "Growth and Development of Cultured Plant Cells," in *The Realm of Life,* vol. 18. New York: McGraw-Hill Book Company, 1972.

Sullivan, W., "Frozen Mouse Embryos Grow in Foster Mother," *New York Times,* August 15, 1972.

Taylor, G. R., *The Biological Time Bomb.* New York: Mentor Press, 1968.

Thomas, L., "On Cloning a Human Being," *New England Journal of Medicine* 291 (1974):1296–1297.

Toffler, A., *Future Shock.* New York: Random House, 1970.

Vasil, V., and Hildebrandt, A. C., "Differentiation of Tobacco Plants from Single Isolated Cells in Microcultures," *Science* 150 (1965):889–892.

Veatch, R. M., "Ethical Issues in Genetics," *Progress in Medical Genetics* 10 (1974):223–264.

Veomett, G., and Prescott, D. M., "Reconstruction of Cultured Mammalian Cells from Nuclear and Cytoplasmic Parts," *Methods in Cell Biology* 13 (1976):7–14.

Wade, N., "Genetics: Conference Sets Strict Limits to Replace Moratorium," *Science* 187 (1975):931–935.

———, "Dicing with Nature: Three Narrow Escapes," *Science* 195 (1977):378.

Wald, G., "The Case Against Genetic Engineering," *Current,* November 1976.

Watson, J. D., "The Future of Asexual Reproduction," *Intellectual Digest* 2 (1971):69–74.

———, "Moving Toward the Clonal Man—Is This What We Want?" *Atlantic Monthly,* May 1971.

————, and Crick, F. H. C., "Molecular Structure of Nucleic Acids," *Nature*, April 25, 1953.

Whitten, W. K., "Nutrient Requirements for the Culture of Preimplantational Embryos *in Vitro*," in G. Raspe, ed., *Shering Symposium on Intrinsic and Extrinsic Factors in Early Mammalian Development, Advances in Biosciences, 6.* Elmsford, N.Y.: Pergamon Press/Viewings in the Press, 1970.

Whittingham, D. G., "Survival of Mouse Embryos after Freezing and Thawing," *Nature* 233 (1971).

Wolstenholme, G. E. W., and Fitzsimons, D. W., eds, *Law and Ethics of A.I.D. and Embryo Transfer, Ciba Foundation Symposium 17* (new series). Amsterdam: Associated Scientific Publishers, 1973.

Yunker, B., "The Baby Bubble," *Ladies' Home Journal,* September 1969.

INDEX

235